Designing, Adapting, Strategizing in Online Education

Also from Westphalia Press

westphaliapress.org

Designing, Adapting, Strategizing in Online Education

Volume 2, Number 1 of
Internet Learning

Edited and Introduced by
Phil Ice

WESTPHALIA PRESS
An imprint of Policy Studies Organization

Designing, Adapting, Strategizing in Online Education
Volume 2, Number 1 of *Internet Learning*

Westphalia Press
An imprint of Policy Studies Organization
1527 New Hampshire Ave., N.W.
Washington, D.C. 20036
dgutierrezs@ipsonet.org

ISBN-13: 978-0918592408
ISBN-10: 0918592402

Updated material and comments on this edition
can be found at the Westphalia Press website:
www.westphaliapress.org

Table of Contents
The following overview of articles was provided by:
Ioana Andreea Stanescu, Project Manager at Advanced Distributed Learning Romania

eLearning Perspectives

Ion Roceanu
> This research focuses on the architecture of an ADL environment, highlighting the learners' need and the dimensions that enable "real" learning to take place. The author stresses on the importance of placing learners at the centre of the process and of considering behaviours, attitudes, aptitudes, timing, and knowledge acquisition when using ADL systems. The main steps required to develop an integrated ADL system are detailed, with focus on military educational resources. These steps concern the identification of the educational types and of the educational resources, the selection of the target groups, the investigation of educational needs and of the basic requirements for learners and tutors. An evaluation of the requirements, carried out in real time on three eLearning platforms, is provided considering three key elements: objectives, actions and techniques. The study highlights the importance of correlating the system parameters with real needs and the user profiles.

Quang Dung Pham & Magda Florea Adina
> This research addresses the lack of adaptation in existing educational systems, where individual learners are delivered the same learning content, and focuses on personalized e-learning systems using both ontology technology and intelligent agents. The paper present a domain ontology that is suitable for adaptive e-learning environments and that describes the learning styles of the learners. A multi-agent e-learning system – POLCA - that aims to provide learners with appropriate learning objects according to their learning styles has been developed in order to be able to assess the efficiency of learning process. The authors provide extensive and consistent theoretical and practical information on learning styles, ontology design, learning object labelling, learning style estimation, and the architecture of adaptive multi-agent e-learning systems. The case study the authors have carried out on 91% has revealed that the participating students evaluated that the system dynamic adaptation is good and very good.

Carmen Holotescu & Gabriela Grosseck
> Social media plays a significant role in higher education. The authors analyse how the academic staff use social media platforms, with the purpose of formulating specific measures that fundament a better performance and a more efficient employment of social environments. The paper presents the results of an online questionnaire that has identified the main reasons people in educational environments use social platforms, what platforms are most popular, how users interact, what are the main activities people use social media for, or the contextual conditions in which scholars use social media. The authors have highlighted the advantages and the disadvantages that have been expressed by the participants, and have discussed the necessity to establish an institution-wide Social Media Observer that strengthens university policies related to social media at the level of the higher education institution

Serious Games: Emerging Opportunities for Education

> *The authors approach interoperability as a key factor that influences the efficiency of serious games development. They explain the need for adopting interoperability perspectives as promoters for efficiency, responsiveness, and cost reduction, and provide evidence why interoperability fails not only due to technical issues, but also to people's inability to apply best practices. The Serious Games Multidimensional Interoperability Framework (SG-MIF) is advanced as a solution that enables SG develops build customizable interoperability scenarios that empower them to consider the advantages and the disadvantages of various alternative solutions they might choose to employ, and thus enhance their decision-making process. Three case studies are presented: Serious Games and standards: SCORM and LOM; Interoperability between games and Learning Management Systems; and Interoperability between game components.*

> *The rapid development of the serious games field has opened new areas of research. This paper present the serious games curriculum integration issues from two perspectives: one of the teachers that employs serious games in educational contexts and the other one of the researchers that analysis the pedagogical state of the art. The study concerns the expertise of experts from Romania, Italy and Spain, and it synthetize the main aspects to be considered within curriculum integration. The impact of curriculum integration on learners and their skills is analysed. The practitioners' point of view considers time management, pedagogy, and outcomes. The authors highlight the need for more direct and mutual collaboration of researchers and teachers in all the four basic stages of a field experiment: conception, design, enactment, and evaluation. The main challenges and barriers to curriculum integration are presented, and the need of a strong collaboration between teachers and researchers is proposed in order to be able to apply the optimal solutions*

> *The authors present the Volhov serious game application used for anti-aircraft missile training and related strategies. The application allows users to train based on several pre-set cases for specific aerial targets. Through specific e-learning strategies the user capability and experience are enriched altogether with specific automatism directly related to the field of anti-aircraft missile training. The paper provides valuable information on the development process, the technical performance of this system, and the training sessions. The main modules, features and parameters are described, and practical step-by-step details on how to operate the system are provided*

> *The authors approach time as an important factor in management education, where students need to be able to manage their time efficiently in order to obtain real benefits from their learning process. The paper presents a case study that focuses on Time Perspective (TP) in relation to learning performance, using the MetaVals serious game as a financial tool. MetaVals is a computer-based Serious Game designed by ESADE in the context of the FP7 Network of Excellence Games and Learning Alliance (GaLA). Master students participating in this case study were engaged in an introductory finance course in ESADE Law and Business School. The data analysis has shown that there is no significant relation between TP and the Game Performance, neither at individual level, nor during collaborative*

Introduction to the Special Issue

As American Public University System's Vice President of Research and Development, I attend over 20 conferences per year, with a twofold purpose. First, I speak on behalf of our University to showcase internal research and innovation. Second, and perhaps most important, I attend these conferences to acquire an understanding of what others are doing in the field of online learning. As the world becomes increasingly interconnected it is imperative that we understand how others are approaching the field and learn from their endeavors.

To provide a truly global education to students, we must move outside of the North American context and understand how the rest of the world views learning and the pathways they are exploring to optimize the student experience. As such, I typically attend six to eight international conferences per year, with as a diverse as possible geographic footprint. In fulfilling this goal I discovered the International Scientific Conference on eLearning and Software for Education (eLSE) in 2009. After having two proposals accepted, my friend and colleague, Jennifer Richardson, an Associate Professor at Purdue University, and I travelled to Bucharest, Romania to present our papers.

Having not previously read any works from Romanian scholars, I was quite interested to discover what the conference would hold. While very small in numbers, I discovered that the conference was extremely rich in knowledge. In fact, I was astounded at the quality of the work that was on display; particularly the spectacularly high-quality gaming and simulation work that was being used or in development.

Another factor that makes eLSE unique can be found in the conference's mission statement: *"The purpose of the annual international scientific conference on 'eLearning and software for education' is to enable the academia, research and corporate entities to boost the potential of the technology enhanced learning environments, by providing a forum for exchange of ideas, research outcomes, business case and technical achievements."*

All too frequently I have witnessed a division between academic and corporate entities, with one side giving little attention to the contribution that one makes to the other. At eLSE all entities are viewed as valuable partners in creating rich, dynamic experiences for learners. In addition, there is a substantial military presence at the conference, helping create another bridge between the private, public, and governmental sectors. From my perspective there is much that other organizations and conferences could learn from adoption of this philosophical framework.

Since my first eLSE, I have looked forward to the next year's conference as much as any event that is on my calendar, and watched attendance grow dramatically year after year. Conversely, the eLSE organizers have been extraordinarily welcoming to me, by encouraging me to return each year and inviting me to join the Scientific Committee; an honor that I gladly accepted. Thus, it is with great pleasure that I have organized this special issue to introduce readers to the eLSE conference. In this publication, seven articles, from the 2012 conference, are presented in two different categories, with a forward by my friends and distinguished Romanian colleagues. It is my intent that another issue will be produced following this year's conference and hope that at least some readers will be able to join us in Bucharest on April 25th and 26th of 2013. http://www.elseconference.eu

Sincerely,
Phil Ice, Ed.D., Editor

VP, Research & Development, American Public University System
The International Scientific Conference eLearning and Software for Education | eLSE

The main scope of the annual **International Scientific Conference "eLearning and Software for Education" - eLSE** is to facilitate the communication and collaboration between national and international academic and business entities, in order to stimulate the research and the development potential of technology enhanced learning environments. The conference provides opportunities to exchange ideas, research outcomes, business cases and technical developments.

The conference proceedings are:
- listed in the Thomson Reuters ISI Web of Science,
- indexed by the Central and Eastern European Online Library,
- listed in the ProQuest database
- listed in the EBSCO database of Conference Proceedings

Over the years, the conference has benefited of the active support of the Advanced Distributed Learning (ADL) Initiative, of universities and companies. Among the keynote speakers invited in the last three years, we mention: Prof. Dr. Teodor Frunzeti, Rector of the Carol I National University of Defence, Mr. Tom Archibald (ADL), Mr. Robert Wisher (ADL), Mr. Jonathan Poltrack (ADL), Mr. Joe Camacho (Joint Knowledge Online – JKO).

The main supporting organisations are: Carol I National Defence University, the Romania Advanced Distributed Learning Partnership Laboratory, The University of Bucharest, and Politehnica University of Timisoara. Among the private companies that have been actively involved in supporting the conference and that have been present at the event, as well as media partners, we mention: Siveco, IBM, Softwin, Advanced Technology Systems, Avitech, Insoft, Computerland, IDG.

The first edition of the International Scientific Conference eLearning and Software for Education was organized in 2005, by the Carol I National Defence University, under the coordination of Prof. Dr. Ion Roceanu and has focused on the applicability of eLearning tools in military education. The organisation of the second edition has been carried out by the new Advanced Distributed Learning Department established at the university. Two volumes with the conference papers have been published.

The first website dedicated exclusively to the eLSE conference was developed in 2007. It has been a success not only because it provided information on the objectives of the conference, the scientific committee, the registration process, the conference program, the registered papers enabled the online registration and management of participants, but also it enabled the online registration and management of the participants.

In 2008, the conference proceedings were indexed in international databases, and the number of registered participants has increased to over 100. Starting with 2009, an anti-plagiarism application has been installed on the conference online platform and all the conference papers have been checked to ensure their originality. In 2010, there were over 300 registered participants. The conference has been organised at the National Military Circle in Bucharest at the same time with a military workshop, where the conference participants could also attend. In 2011, due to the large number of papers, the conference program has included three parallel sessions.

The 2012 edition of the conference focused on the following main topics:
1. Management strategies and policies
2. Pedagogy and psycho-pedagogy in new learning environments
3. Computer science and new support technologies in learning
4. Serious games in theory and practice
5. Corporate eLearning and training
6. E- Content / Instructional design

The participants came from over 22 countries, and from over 20 cities in Romania. 12 military organisations, 32 international universities, 33 national universities, and 8 companies were present. 199 papers have been presented at the conference.

The conference aimed to highlight the importance of standardisation in eLearning environments and Mr. Jonathan Poltrack from ADL has been invited to present the opportunities of implementing standardisation solutions, with focus on the Sharable Content Object Reference Model (SCORM). For the first time, the conference has included a section dedicated exclusively to serious games.

eLSE   http://www.elseconference.eu

Designing The Military Advanced Distributed Learning System

Ion Roceanu
Advanced Distributed Learning Department, "Carol I" National Defense University

Abstract
An Advanced Distributed Learning (ADL) environment needs to be extended beyond technical drivers to pedagogical and organizational dimensions that focus on the interaction between the learner and the learning environment. In fact, effective e-learning resources can not only be used to complement face-to-face education or replace the classroom for distance education, but can also facilitate the integration of student interaction and real-world scenarios into the learning process. The use of highly interactive and virtual resources can support authentic learning where students can relate to and experience real world contexts in their learning. This was the main road that helped us in developing one of the most powerful ADL systems in the military education institution. This experience could be extended to the civilian and corporate realms to serve as a guide in the tentative design of such a system.

KEY WORDS: *e-learning, LMS, advanced distributed learning, learning architecture*

I. General ADL Environment Architecture

The Advanced Distributed Learning (ADL) Initiative was launched in 1997 as a visible commitment to incorporate into practice the benefits of technology-based instruction, generally referred to as e-learning. The goal of the ADL Initiative is to ensure access to high-quality education, training, and job support, tailored to individual needs and delivered on-demand anytime and anywhere.

The lofty vision of ADL required a new approach to doing business; one based not on a belief of "build it and they will come" but on a belief that sustainable advances in e-learning could be best achieved through cooperative efforts.

The general ADL architecture framework is as shown in Figure 1. To define a business paradigm of the ADL system, we could take into consideration the principles of the Enterprise Architecture Framework. Based on it, it becomes necessary to define the components: Business Reference Model (BRM), Service Component Reference Model (SCRM) and Performance Reference Model (PRM). These will be defined by compiling the results from the first three stages which are described in chapter III. Consequently, the Business Reference Model (BRM) will underline the correct relationship between the organizational requirements of the military educational system (objectives and specific resources) as well as the technical structure of the basic components of an ADL Environment (learning content management system, content repository, educational resources, students, tutors, subject matter experts, instructional designer, etc.)

The Service Component Reference Model (SCRM) is focused on bridging gaps between students' and tutors' requirements in relation to the psychopedagogical tools in use.

Performance Reference Model (PRM) targets the continuous assessment tools in performing the educational process based on integrated ADL system components. Basically, what this segment is meant for is identifying those tools by means of which an ADL environment is perfectly adjustable to real needs in the educational system by assessing results and turning these into requirements.

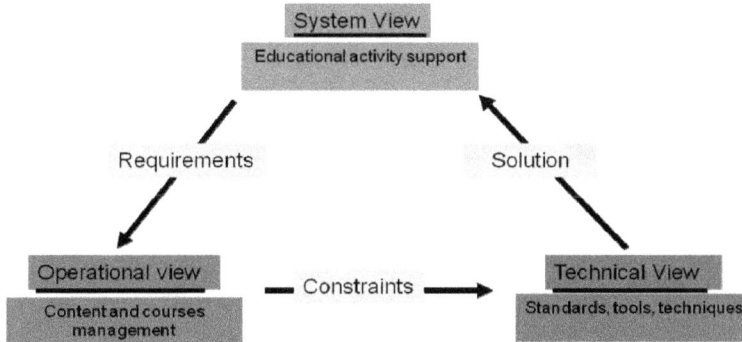

Figure 1. The general ADL architecture view

II. What Do Learners Within An ADL Environment Need?

Many of the e-learning systems developed thus far incorporate features requested by technologists and teachers and are designed from their point of view. This lack of learners' perspective has led to many Learning Management Systems (LMS) being rejected or being scored low in reviews by the learners.

As in any classroom situation, but especially in distance learning areas, there is conflict between teachers and learners regarding the most appropriate and effective didactical tools used. There is an emphasis on continuous development of existing methods of teaching in the traditional classroom situation in order to maintain the learners' attention, meet their learning needs, and for them to ultimately achieve the intended outcomes. If this approach is not given equal importance within distance learning activities, some individuals will learn little that can be effectively transferred to the workplace, or even worse, may choose not to complete the course at all.

In military training activities, the situation is made even more difficult than that of civilian training for at least three reasons:

1. The group of learners is very heterogeneous regarding the level of education, background, age, specific professional skills, motivation, IT knowledge, and so on.
2. The instructors come from different domains, with or without didactical experience.
3. There is a large palette of content and training objectives and no standard time for completion of courses (from 1 hour to months).

The e-learning system used to provide distance learning courses must address, in the first instance, the type of features required by the main actors of the learning activity, the learners themselves.

The ADL concept is a new and very productive one from a technical point of view but there are still many areas for improvement, especially in the didactical aspects. One of the sensitive subjects is represented by the methods used to attract the learners and to meet their expectations regarding the distance training environment.

2.1. Putting the learners at the center of the process

Of course any training or learning intervention must be driven by the organization's objectives. However, the intervention will be much more successful if the learner is put at the center of the process for achieving this. Learning is a journey, which takes time and patience and, most importantly, self-motivation. It is vital that the learning environment is such that it supports students to learn in the best way possible for them as individuals and thus encourage motivation and the stimulus to learn.

Adults have a vast repertoire of experience that can and should be drawn on in order for them to learn effectively and to enjoy the learning experience. It makes sense then, that the existing skills and experience of individuals using the ADL environment should be identified prior to commencement of any learning and that the systems and tools used should be able to support these. A single teaching technique is not suitable for all learners and effective, fulfilling, and successful learning will not take place unless individuals are provided with a variety of different techniques, approaches, and environments.

Hartley identified six key cognitive principles of learning which relate to inferences, expectations, and making connections:

a) Instructions should be well organized.
b) Instructions are clearly structured.
c) Perceptual features are important.
d) Prior knowledge is important.
e) Individual differences are equally important to learning. Each individual has different approaches of learning and cognitive styles.
f) Students require cognitive feedback that provides them with information regarding success or failure.

The approach to ADL thus far has been to look at the capabilities of the system and "fit" the learners into this, as opposed to putting the learners at the center of the process. Hartley's key principles are unlikely to be fully addressed using this approach.

As mentioned earlier, motivation is a key factor for successful learning and assimilation of knowledge. According to Mexirow's Charter for Andragogy, adult learners learn through both self-directed learning and learner self-direction (personal characteristics). Learner self-direction centers on the learner's desire for assuming the responsibility for learning. Based on the premise that the ADL environment does not involve a teacher, it becomes even more important that the design and delivery capabilities of the system, tools, and courses are able to motivate a learner to assume responsibility for their learning. An example of this is having a system which provides students with an individual learning pathway tailored to their identified needs and a comprehensive method of recording results and giving feedback.

2.2. Behavior, attitudes, aptitudes, and knowledge

The most commonly used model for evaluating the learning process is Kirkpatrick's which identifies four levels that need to be measured.

• Reaction of the student—what they thought and felt about the training.
• Learning—the resulting increase in knowledge or capability.

- Behavior—the extent of behavior and capability improvement and implementation/application.
- Results—the effects on the business or environment resulting from the trainee's performance.

Many educators believe that "real" learning takes place when there has been a change in behavior and attitude, though this level can be difficult to measure. However, what is clear is that in order to evaluate the success of any learning intervention, there needs to be a Training Needs Analysis (TNA) carried out prior to training being developed or undertaken. This includes planning and setting up the evaluation process at the outset. In this instance, the TNA is not just to identify gaps in knowledge and skills, but to ascertain the needs of the students within the ADL environment itself. The TNA must include questions around cultural- and country-based differences and similarities as well as those of individual students. It is not good practice, or cost effective, to simply wait to receive the results of evaluation before attempting to "get the system right the first time". Questions such as "Why did/do some learners achieve better results than others who have experienced the same programme" should be asked prior to the course ever being rolled out. This means carefully looking at the cohort of learners, their previous experiences, their existing skills, and their needs and building the system of delivery around this knowledge. Often, the lack of success of a course and its learners is due to the lack of a TNA being carried out.

Another element that needs to be considered is timing; there is considerable evidence that if a new skill or behavior is not used within a relatively short time, learning degrades very rapidly. The ADL system cannot be used in isolation or instead of "on the job" training.

According to the Johnson O'Connor Research Foundation, "aptitudes are natural talents, special abilities for doing, or learning to do, certain kinds of things easily and quickly. They have little to do with knowledge or culture, or education, or even interests." An individual will have their own intellectual ability to learn material sufficiently in order to perform their job role. However, the delivery mode or system used must be able to meet their specific style or need otherwise they will be less successful at learning. This concept is another reason why the learners' needs within the ADL environment should be identified.

The Oxford English Dictionary defines knowledge acquisition as "involving complex cognitive processes: perception, learning, communication, association, and reasoning". The real power of distance learning and ADL is the potential to interact with individuals who have a vast variety of experiences to share with others from their own, and perhaps more significantly, other countries and cultures. This needs to be exploited to its full potential to allow for sharing of knowledge.

In the first instance, this project will seek to ultimately bring about the changes identified above, through an in-depth investigation, analysis, and evaluation of the learners' needs with regard to the system and tools available. Instead of making assumptions as to what the learner needs, or to simply do what the system is capable of doing, it is necessary to ascertain from the learners themselves what they require.

**III. Steps In Developing An Integrated ADL System By Leveraging The Learners`
Perspectives**

3.1. Learners' and tutors' profile
We should bear in mind that adults are very different in the way they acquire
knowledge and that a wide array of factors are to be considered as influential in their
attitude.
   a) Identifying types of users within the military educational environment: Different
       groups of users within the military will be clearly portrayed by using pre-set
       assessment criteria, such as age, educational background, skill, competences, and
       learning styles. It is important to know the array of differences and similarities of
       learners. We need to make the educational process a student-centered approach,
       where individuals take responsibility for their own resources and make decisions
       about how and when to progress to the next stage. This will make the learner
       process more effective and efficient leading to greater achievement. We need to
       know who the learners are, what their existing skills and educational background
       are, and what further assistance they might need. Likewise, tutors need to clearly
       identify the best training approach required based on the information collected.
   b) Identifying educational resources within the military educational systems: A
       relationship between student's profiles and educational resources will be traced.
       The Honey and Mumford learning styles will be considered (activists, reflectors,
       theorists, and pragmatists) as well as the kinesthetic, visual, and auditory types
       of learners. All this should be related to the very type of resource in focus for the
       intended purpose.
   c) Selecting the target groups with a view to undergo scientific research on: This will
       be performed by means of surveys (questionnaires created for this very purpose).

**3.2. Investigating educational needs and basic requirements for learners and tutors
with a view to military systems educational resources**
In order to gather information on military learners' needs to later relate it to the
tutors involved, a questionnaire needs to be designed. Consequently, the following issues
are to be considered:
   –   The would-be learners' assessment of their previous experience, theoretical
       approach, and educational background. The purpose at this stage is to see if
       the would-be learners translate their previous experience within different
       environments to the military field. If the outcome is positive, then it is an
       important issue to be considered at the time of Instructional Design.
   –   The extent to which ADL is a well-known and understood concept in itself.
   –   A beginner's expectations in the field, in terms of technicality.
       An in-depth analysis of the gathered information will be performed in relation to
initial requirements.

**3.3. An evaluation–investigation session for preferences/requirements in real-time
situations on existing e-learning components (LMS, content, methods, systems)**
By means of these activities, an identification of the expressed options and a correlation
with the concept of putting this into practice within already developed systems is in
focus. This is a mandatory step in order to remove those requirements which are not
sustained by practical reasons for accomplishment. Three different e-learning platforms
will be used, with different educational content, on three targeted groups previously

selected. The content will differ both in form (text, multimedia, dynamics, interactivity), standardization (SCORM 1.2, 2004, AICC) as well as providing method (self-paced, linear, sequential). The testing session will be performed on mixed groups (students/tutors) but also performed separately. The table below gives some instructions for accomplishing these steps.

Objectives	Actions	Techniques
Defining and developing the measurement model	- Definition and organization of methods, algorithms and other aspects of measurement of ADL Environment Enterprise Architecture; - Segregation of educational processes (e.g. studies, courses, lectures, others) to properly survey and evaluate.	- Surveys/questionnaires/organization analysis; - Analysis of important factors related to user profiles, latitude of influence of this factors; - Analysis of known learning evaluation models; - Segregation and definition of educational processes.
Learner's and tutor's profile	- Comparison of content-specific learning objectives; - Define the learner profiles in connection with the content and objectives; - Define the tutor profile; - Define the target groups.	- Analysis of curricula or training programs: National Defense Institutes; NATO and PfP Training centers and schools, other participants.
Theoretical learner's and tutor's needs	Questionnaires	Applied to each target group: newbie e-learning user; experienced learners; tutors; 5–6 countries
Prove the theoretical needs in the real existing LMS	Compare the "blind" needs with the behavior in the real existing LMS. Compare evaluation tools for every LMS platform	Test the ADL system based on three different LMSs in three different conditions; Ro ADL lab will manage the technical issues
ADL Environment Enterprise Architecture	Define the BRM; SCRM and PRM	Leverage the results of the steps above

IV. Conclusions

Designing of an ADL system requires a thorough analysis of the organization and its training needs. On the other hand, it takes a long-term vision so that the system is designed to withstand time, both from a technical standpoint and from the results in training. In this article the author attempted to highlight just one important aspect of the problem - that of system parameters correlating with real needs and profiles of beneficiaries to avoid pre-requisite rejection. Due to rapid developments in technology, technical components of an ADL can be easily solved with minor adjustments, but the systemic component is exclusively operational and organization-specific and requires special attention. In addition, greater attention must be shown regarding how to build training content and how pedagogical tools are used to develop appropriate educational content specific training objectives.

About the Author

Ion Roceanu serves as Director of the Advanced Distributed Learning Department, at the "Carol I" National Defense University in Bucharest, Romania.

References

Advanced Distributed Learning. 2004. *Sharable Content Reference Model (SCORM).* Third Edition, Alexandria, VA: ADL, US Government.

Fletcher, J.D. 1997. "What Have We Learned about Computer Based Instruction in Military Training?" In *Virtual Reality Training`s Future*?, eds. R.J. Seidel, and P.R. Chatelier. New York: Plenum Publishing.

Roceanu, Ion, and Alexandra Toedt. 2008. "Managing the Information Deliver the Knowledge. Steps in Developing the Digital Content,: eLSE Conference, Bucharest.

Roceanu, Ion, and Geir Isaksen. 2010. ADL – European Perspective, in the Learning on Demand, section IV.

Wisher, Robert A., and Badrul Kahn. 2010. Learning on Demand. ADL and the Future of e-Learning.

Adaptation To Learners' Learning Styles In A Multi-Agent E-Learning System

Pham Quang Dung and **Adina Magda Florea**
Politehnica University of Bucharest Bucharest, Romania

Abstract

Adaptation to learners' learning styles can help education systems improve learning efficiency and effectiveness. This research orientation has been studied by many researchers lately, but most of the existing education systems lack adaptation in which every learner is delivered the same learning content. Moreover, many researchers concluded that it is worth applying automatic identification of learning style because of its advantages in precision and time savings. In our study, we concentrate on two main technologies to implement adaptation in education systems: semantic web and intelligent agents. Using ontology with the Semantic Web services makes it faster and more convenient to query and retrieve educational materials. Intelligent agents can provide the learners with personal assistants to carry out learning activities according to their learning styles and knowledge level. In this paper, we present a domain ontology that is suitable for adaptive e-learning environments. The ontology describes the learning objects that compose a course as well as the learners and their learning styles. We also present a multi-agent e-learning system that supports pre-defining and re-examining students' learning styles during the course for better personalization. In the system, the learning style of each learner can be identified automatically and dynamically. We used a new literature-based method that uses learners' behaviors on learning objects as indicators for this task. The evaluation showed a high precision in detecting learning styles and in delivering learning materials. Together with the mentioned benefits, this result indicates that our e-learning system is capable of wide use.

KEY WORDS: *adaptation, semantic web, ontology, personalized, multi-agent, e-learning system.*

I. Introduction

Nowadays, the combination of education and the web leads us to web-based education (WBE) that has become a very important branch of educational technology. In WBE, organization and the access to learning objects (LOs) are important matters. Several standards of LO metadata have been used such as IEEE LOM, SCORM, Dublin-Core. Metadata provides better representation and understanding of learning content, and enables people to transform, share, and reuse learning content. However, the metadata is not enough; it is lacking reasoning capability and machine processing ability (Wang, Fang, and Fan 2008).

By putting WBE in the context of semantic web, we have a new generation of WBE, or semantic web-based education (SWBE). The use of semantic web and web intelligence makes WBE more effective and more appealing to learners, teachers, and authors alike (Devedzic 2006). Ontology is considered as the key concept in semantic web. It represents domain knowledge by defining terminology, concepts, relations, and hierarchies in a machine-readable form. It also makes web-based knowledge easier in processing, sharing, and reusing. The ontological description of LOs can overcome disadvantages when using other representations. Therefore ontology-based learning systems are becoming more common day by day.

Personalization in education is also one of the hottest research and development topics currently. In this context, each learner has his own learning style that indicates how he learns most effectively. Several well-known learning style models are proposed by Myers-Briggs, Kolb, and Felder-Silverman. Personalized e-learning systems allow students to learn by themselves so that it would improve learning effects and overcome the disadvantage of traditional classroom teaching (Min and Lei 2008). Besides ontology technology, artificial intelligent agents can be used to improve personalization in learning systems by tracking learners' activities during the course to estimate their learning style and providing them appropriate LOs.

Our research concentrates on personalized e-learning systems using both ontology technology and intelligent agents. We propose a domain ontology aimed to support personalized online learning.

The ontology describes the learning material that composes a course in terms of both learning resources and acquired knowledge, as well as the learners and their learning styles. The acquired knowledge is structured along competencies and abilities acquired, mapped to concepts and learning resources. A multi-agent e-learning system that can provide learners with appropriate learning objects according to their learning styles was developed in an attempt to assess the efficiency of the learning process.

The rest of this paper is organized as follows: Section II introduces related work including learning object and learning style. Section III presents materials and methodology. In Section IV, we discuss our results, and Section V draws on conclusions and future work.

II. Related Work

2.1. Learning object

The expression "learning object" is one of the most cited terms in e-learning literature. However, this term is not cited within relevant terminological reference sources, such as the Oxford English Dictionary, the Merriam-Webster Dictionary, or the WordReference website. About this problem, McGreal (2004), in his study on LOs definitions, highlighted that there are five types of definitions most used:

i.   "anything and everything;
ii. anything digital, whether it has an educational purpose or not;
iii. anything that has an educational purpose;
iv. only digital objects that have a formal educational purpose;
v.   only digital objects that are marked in a specific way for educational purpose."

Some research has been carried out with the aim of investigating the LO's domain from a formal ontological perspective, for example the study conducted by Sicilia et al. (2005), starting from the previously cited research of McGreal, proposed an original ontological schema as an investigative tool for learning objects description. Their results show that an LO can be ontologically defined as "any physical object which is purposively designed and developed in order to support someone to reach at least one learning objective".

2.2. Learning styles

2.2.1. Learning style concepts

Some authors have proposed different definitions for learning style. For example, in (Riding and Rayner 1998) learning style is described as an expression of individuality, including qualities, activities, or behavior sustained over a period of time. In educational psychology, style has been identified and recognized as a key construct for describing individual differences in the context of learning.

Keefe (1979) defines learning styles as "cognitive characteristics, affective and psychological behaviors that serve as relatively stable indicators of how learners perceive, interact with and respond to the learning environment."

James and Gardner (1995) define learning style as the "complex manner in which, and conditions under which, learners most efficiently and most effectively perceive, process, store, a n d recall what they are attempting to learn" (p. 20). Merriam and Caffarella (1991) present Smith's definition of learning style, which is popular in adult education, as the "individual's characteristic way of processing information, feeling, and behaving in learning situations" (p. 176) (James et al. 1998).

2.2.2. Felder-Silverman learning style model

Several well-known learning style models were proposed. In our research, we concentrate on the Felder-Silverman model (Felder 1988) because the authors provide the questionnaire and a completed guide to use it. Moreover, this model has been proven to be effective in many adaptive learning systems (Hong and Kinshuk; Peña, Marzo, and de la Rosa 2005; Zywno).

The learning style model was developed by Richard Felder and Linda Silverman in 1988. It focuses specifically on aspects of learning styles of engineering students. Three years later, a corresponding psychometric assessment instrument, Felder–Soloman's Index of Learning Styles (ILS), was developed.

Their model permits classification of students into four categories, Sensory/Intuitive, Visual/Verbal, Active/Reflective, and Sequential/Global. The dimensions Sensory/Intuitive and Visual/Verbal refer to the mechanisms of perceiving information. The dimensions Active/Reflective and Sequential/Global are concerned with processing and transforming information into understanding (Soloman and Felder).

The ILS instrument is composed of 44 questions, 11 for each of the four dimensions previously described. This questionnaire can be easily completed through the web (Soloman and Felder) and provide scores as 11A, 9A, 7A, 5A, 3A, 1A, 1B, 3B, 5B, 7B, 9B, or 11B for each of the four dimensions. The score obtained by the student can be:

- 1–3, meaning that the student is fairly well balanced on the two dimensions of that scale;
- 5–7, meaning he has a moderate preference for one dimension of the scale and will learn more easily in a teaching environment that favors that dimension;
- 9–11, meaning that he has a very strong preference for one dimension of the scale and probably has immense difficulty in learning in an environment that does not support that preference.

The letters "A" and "B" refer to one pole of each dimension.

III. Materials and Methodology

3.1. Ontology design

The representation of learning objects using metadata is not good enough because of the lack of machine processing ability and reasoning capability. With the development of semantic web and ontology, all these problems can be overcome because ontology is good at reasoning and is machine-readable. The use of ontology to represent learning objects enable different education applications to share and reuse the same educational contents. Furthermore, the machine-readable ability of ontology enhances the speed of query processes and the accuracy of the responded results. Hence, learners can have the learning objects they need quickly and they can be more reliable.

José M. Gascueña, Antonio Fernández-Caballero, and Pascual González (2006) proposed a domain ontology for personalized e-learning in education systems. They considered two characteristics that describe each educational resource which are: (1) the most appropriate learning style and (2) the most satisfactory hardware and software features of the used device. Starting from the ontology proposed by Gascuena, Fernandez-Caballero, and Gonzalez (2006), our work concentrates on developing an e-learning system that works well on PCs with a web browser, not on limited memory and screen size devices such as PDAs.

3.2. Learning objects labeling

Each learning object is labeled with one subtype of any element in the set of 16 types of combinations from four categories mentioned in Section 2.2.3. For example, learning object 1 is labeled as ActiveSensingVisualSequential, while learning object 2's label is Visual only.

Based on the theoretical descriptions about learning style characteristics of Felder–Soloman, and on the practical research of Graf, Kinshuk, and Liu (2008), Hong and Kinshuk, and Popescu, Trigano, and Badica (2008), the learning objects in the POLCA system are labeled as described in Table 1.

Table 1. Labels of learning objects in POLCA

Active	Reflective	Sensing	Intuitive	Visual	Verbal	Sequential	Global
Self-assessment exercises, multiple-question-guessing exercises	Examples, outlines, summaries, result pages	Examples, explanation, facts, practical material	Definitions, algorithms	Images, graphics, charts, animations, videos	Text, audio	Step-by-step exercises, constrict link pages	Outlines, summaries, all-link pages

3.3. Learning styles estimation

Completing the Felder-Silverman questionnaire when first logging in to the system is an optional choice for each learner. If he takes that entry test at that time, the system can deliver learning materials adaptively for him immediately. Otherwise, the adaptation for the learner will start only from the point when the system identifies his learning style automatically.

We used a literature-based method to estimate learning styles automatically and dynamically. Expected time spent on each learning object, $Time_{expected\_stay}$, is determined. The time that a learner really spent on each learning object, $Time_{spent}$, is recorded. These pieces of time are also the ones calculated for each learning style labeled for the learning objects. For instance, if $\text{Time}_{\text{expected\_stay}}$ of a ReflectiveSensing learning object is 30 ms, then $\text{Time}_{\text{expected\_stay}}$ assigned for Reflective, as well as for Sensing is 30 ms. After a period P, which is passed as a system parameter (for example, six weeks), sums of $\text{Time}_{\text{spent}}$ for each of all the eight learning style elements of the learner is calculated. Then we find out eight respective ratios:

$$RT_{LS\_element} = \frac{\sum \text{Time}_{\text{spent}}}{\sum \text{Time}_{\text{expected\_stay}}}$$

We use the same manner to find out the ratios $RV_{\text{LS\_element}}$ which are concerned with the number of visits aspect. Number of learning objects visited and total of learning objects with respect to each learning style element are counted for in the calculation.

$$RV_{LS\_element} = \frac{\sum \text{LOs}_{\text{visited}}}{\sum \text{LOs}}$$

Finally, we calculate the average ratios:

$$R_{\text{avg}} = (RT + RV)/2$$

Learning styles are then estimated based on the following simple rule:

R_{avg}	LS preference
0–0.3	Weak
0.3–0.7	Moderate
0.7–1	Strong

The mutual results for two learning style elements of the same dimension, which are both strong, are rejected. Obviously, a learner cannot have both strong Active and strong Reflective learning style. One other ability is that R_{avg} for both elements of one dimension are less than 0.3. At the current round of adaptation, we no longer consider this dimension because it is not needed to provide the learner with learning materials that match this part. We will finish this sub-section by showing the learning style of a learner's example result presented in Table 2.

Table 2. An example result of calculated R_{AVG}

	ACT	REF	SNS	INT	VIS	VRB	SEQ	GLO
R_{avg}	0.5	0.6	0.25	0.2	0.8	0.15	0.8	0.9

Applying the rule, we define that the learning style of the learner is moderate Active/Reflective, and strong Visual. In this situation, the pair SEQ/GLO is rejected, and the pair SNS/INT can be ignored.

3.4. Learning objects delivery

Once a learner's model is updated, the system delivers only the learning objects that match his learning style to him. The match can be explained as: Learning objects with learning style LS will match a learner with learning style moderate/strong LS. For the learner in the previous example, he will receive only learning objects, whose learning style labels consist of Active, or Reflective, or Visual.

IV. Results and Discussion

4.1. The ontology

In our ontology, we consider only the learning style characteristic, we have added some classes and properties, and we have modified some relationships to make it more reasonable for real courses. Figure 1 shows the layout of the domain ontology that we developed.

Each course has its objectives including competence, knowledge, and abilities. There is a competence per objective. For example, after taking the Artificial Intelligence (AI) course, learners are able to solve complex problems in AI. There are several pieces of knowledge (concepts) and abilities that will contribute to the achievement of a given competence. Here knowledge, of course, means theoretical angle, and abilities correspond to practical skills. Class Ability was added because of this reason.

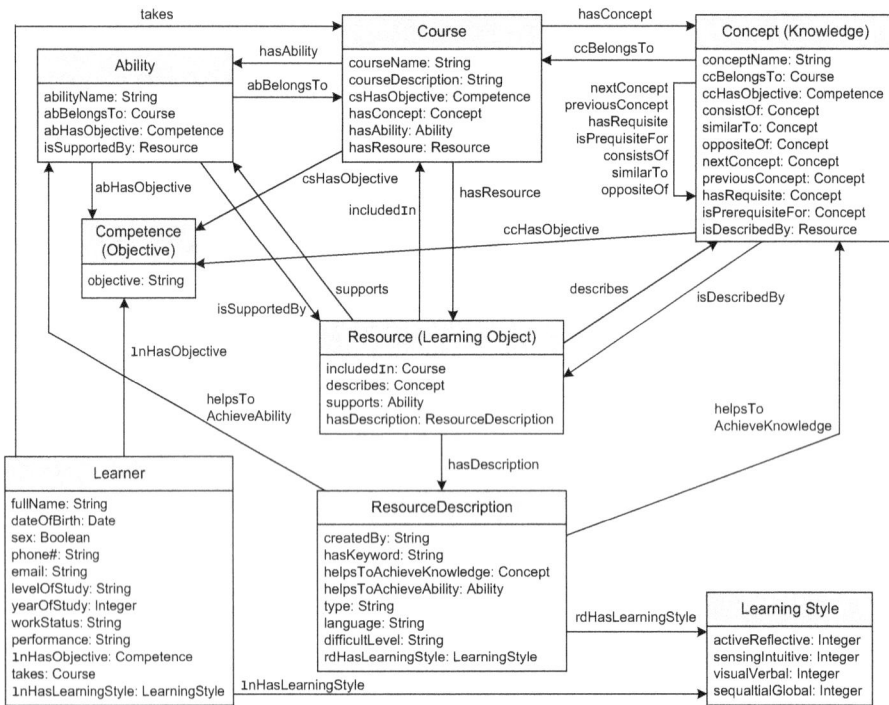

Figure 1. General layout of domain ontology

Like class Concept, class Ability contains abHasObjective property, and isSupportedBy (supports is its inverse) pointing to the set of resources (learning objects) that support the ability.

A resource, or a learning object, can be included in several courses; it can reference several concepts; and it can support several abilities. Class ResourceDescription describes a learning object more clearly. Some added properties are:

(1)   helpsToAchieveKnowledge and helpsToAchieveAbility respectively point to the knowledge and the ability that it helps to achieve.

(2)   type—a learning object can be: one to several PowerPoint slides, one animation that illustrates the concept, one picture or several pictures, one multiple choice exercise, one input text exercise, one programming exercise, one http address, one article, etc.

We first use Felder-Silverman Learning Style Model to identify learners' learning styles for our e-learning system. We assign rdHasLearningStyle property for learning objects so that they can be adaptively delivered to learners.

Class Learner was added because learner is the most important factor of adaptive learning systems. As one can observe, each learner (a) has his name (fullName); (b) has a date of birth (dateOfBirth); (c) is male or female (sex); (d) has a phone number (phone#); (e) has an email (email); (f) is a graduate student or an undergraduate student (levelOfStudy); (g) is in which year of study (yearOfStudy); (h) studies on-campus or off-campus (workStatus); (i) has his performance (performance) that can be excellent, good, average, bad, terrible; (j) has his learning objective (lnHasObjective); (k) has a list of courses that he has to take (takes); and (l) has a learning style (lnHasLearningStyle). This last property together with the same property of the learning object, of course, helps to implement personalization in the learning system.

4.2. POLCA, an adaptive multi-agent e-learning system

4.2.1. System architecture

The e-learning system we have been developing is a multi-agent one, human and artificial agents work together to achieve the personalization and learning tasks. There are two agents that are responsible for personalizing in the system: the learning style monitoring agent and the adaptive content agent. During the courses each learner takes, the first agent monitors his learning activities in order to re-estimate his learning style and give him advice if it is different from his recorded one by a test. The second agent, adaptive content agent, decides which learning objects should be delivered to each learner according to his learning style. Figure 2 shows the architecture of the system.

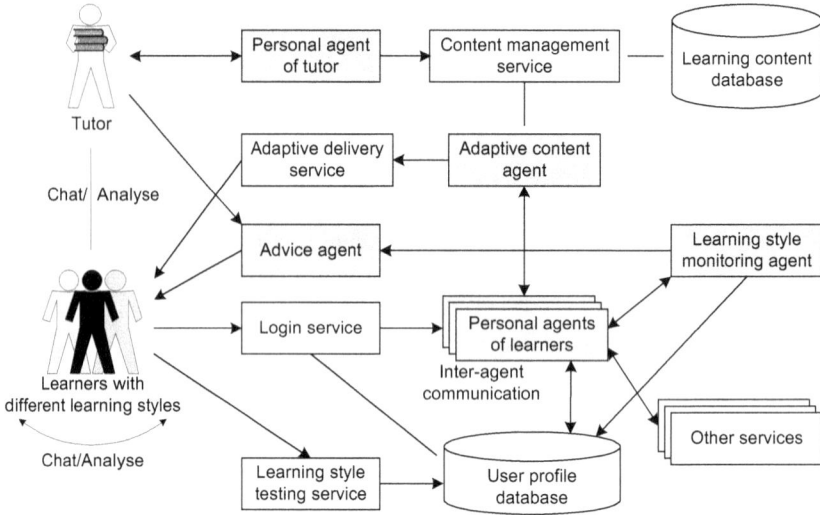

Figure 2. Architecture of adaptive learning style e-learning system
based on intelligent agents and services

4.2.2. System operation

Based on the architecture, a multi-agent e-learning system has been conducted to evaluate the adaptation method mentioned above. Members that can participate in the systems are administrators, teachers, and learners. The learning process starts when a teacher updates his course's learning units, i.e. learning objects.

After being activated by the administrator, a learner can sign in to the system and apply for a new course or navigate through learning units of permitted courses. The learner can choose the way that presents learning units: (1) normal way—all learning units will be shown; (2) adaptive way—only learning units matching his learning styles will be shown.

Student's learning style discovered at the moment is compared with his previous one. If there is no difference, then the adaptation stays the same. Otherwise, the system notices the user and automatically applies adaptation according to his newly detected learning style (Figure 3).

Figure 3. A screen shot from POLCA to which a teacher adds a learning object

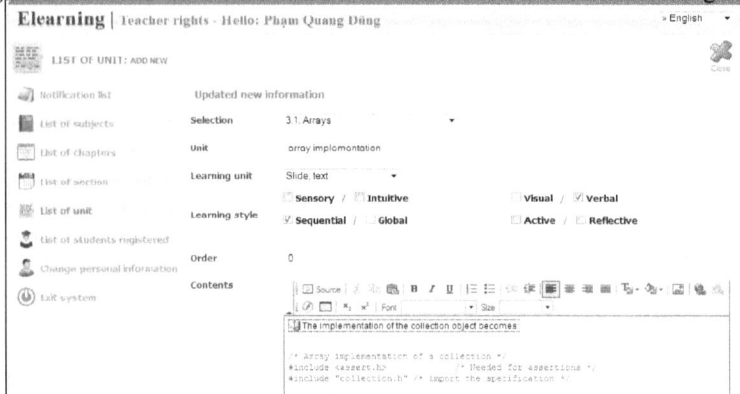

We chose an Artificial Intelligence (AI) course to evaluate our method. The duration for the experiment was nine weeks; that is enough for studying nine sections with 204 learning objects included. The learning objects are sufficient as described above. The parameter P was set to four weeks. Forty-four undergraduate students in the field of Computer Science from Politehnica University of Bucharest participated in the study. They were finally asked to fill the ILS questionnaire and to give feedback about system adaptation. The comparison of learning style detection between our method and the ILS questionnaire is (72.73%, 70.15%, 79.54%, and 65.91%) corresponding to four learning style dimensions Act/Ref, Sen/Int, Vis/Vrb, and Seq/Glo. Regarding the adaptation process, 91% of participating students evaluated that the system dynamic adaptation is good and very good.

V. Conclusions and Future Work

In this paper, we have presented a domain ontology that is suitable for the system mentioned above. The objective and the components of a course are fully described. Students' learning styles are included in the descriptions of both learners and learning objects. This helps adaptive implementation more accurately.

We have also proposed an architecture for building a personalized multi-agent e-learning system. Such a system has been developed. The system uses intelligent agents to re-estimate learners' learning styles and to deliver learning objects fit for each student. One of our future goals is t o implement the system using discussed ontology. Extensive testing is also required in order to firmly validate the proposed system and the efficiency of the approach.

About the Authors

Pham Quang Dung is a PhD candidate at the University "Politechnica" of Bucharest, as well as a lecturer at the Hanoi University of Agriculture in the Information Technology and Computer Science departments.
Adina Magda Florea is Dean of the Faculty of Automatic Control and Computers, and a professor in the Department of Computer Science at the University "Politechnica" of Bucharest.

References

Devedzic, V. 2006. *Semantic Web and Education.* USA: Springer Science-Business Media, LLC.

Felder, R.M. 1988. "Learning and Teaching Styles in Engineering Education." *Journal of Engineering Education* 78 (7): 674-681.

Gascuena, Fernandez-Caballero, and Gonzalez. 2006. "Domain Ontology for Personalized E-Learning in Educational Systems." In *Proceedings of the Sixth International Conference on Advanced Learning Technologies (ICALT'06)*, 456-458.

Hong, H., and Kinshuk. "Adaptation to Student Learning Styles in Web Based Educational Systems." *Proceedings of ED-MEDIA 2004, World Conference on Educational Multimedia, Hypermedia & Telecommunications*, 491-496.

James et al. as cited in Bettina Lankard Brown. (1998). Learning Styles and Vocational Education Practice. ERIC Clearinghouse on Adult, Career, and Vocational Education. [online], Columbus, OH. http://www.calpro-online.org/eric/pab.asp

Keefe, J.W. 1979. "Learning Style: An Overview." In *Student Learning Styles: Diagnosing and Prescribing Programs*. ed. J.W. Keefe. Reston, VA: National Association of Secondary School Principals.

Graf, S., Kinshuk, and T.C. Liu. 2008. *"Identifying Learning Styles in Learning Management Systems by Using Indications from Students' Behaviour."* In *Proceedings of the 8th IEEE International Conference on Advanced Learning Technologies (ICALT'08)*, 482-486.

Min, Wei, and Lei. 2008. "Research of Ontology-based Adaptive Learning System". In *Proceeding of International Symposium on Computational Intelligence and Design (ISCID '08)*, 366-370.

Ontology concept by wikipedia. http://en.wikipedia.org/wiki/Ontology_information_science

Peña, C.I., J.L. Marzo, and J.L. de la Rosa. 2005. "Intelligent Agents to Improve Adaptivity in A Web-Based Learning Environment." *StudFuzz* 178, 141-170.

Popescu, E., P. Trigano, and C. Badica. 2008. *Relations between Learning Style and Learner Behavior in an Educational Hypermedia System: An Exploratory Study.*

In *Proceedings of the 8th IEEE International Conference on Advanced Learning Technologies (ICALT'08)*, 725-726.

Riding, R. and S. Rayner. 1998. *Cognitive Styles and Learning Strategies: Understanding Style Differences in Learning and Behavior*. London: David Fulton Publishers Ltd.

Soloman, B.A., and R.M. Felder. "Index of Learning Styles." http://www.engr.ncsu.edu/learningstyles/ilsweb.html

Wang, X., F. Fang, and L. Fan. 2008. "Ontology-Based Description of Learning Object". In *The 7th International Conference on Web-based Learning (ICWL 2008)*, LNCS 5145, 468-476.

Zywno, Malgorzata S. "A Contribution to Validation of Score Meaning for Felder-Soloman's Index of Learning Styles". In *Proceedings of the 2003 American Society for Engineering Education Annual Conference & Exposition*.

An Empirical Analysis Of The Educational Effects Of Social Media In Universities And Colleges

Carmen Holotescu
Politehnica University Timisoara, Romania

Gabriela Grosseck
West University of Timisoara

Abstract

In an era of fundamental changes in education brought about by virtual worlds and augmented reality, dominated by mobile devices and applications, it is necessary to rethink the academic work environments based on the use of social applications like Facebook, YouTube, or Twitter, in accordance with the skills and learning needs of students. In this context the authors discuss how today's Romanian higher education actors perceive and use social media, trying to find out the answers to questions such as: How faculty members use social media as reflective and collaborative teaching and learning tools, also for research and professional development? Which are the potential benefits, challenges, and disadvantages in using social media in universities? Is there a need for training the educational actors in this topic? Thus in order to shed light on the research issues, we have developed and applied an online questionnaire for scholars from different universities and colleges from Romania. Although our findings revealed an increasing use of social media by educational actors for the time being, only a few universities have adopted coherent strategies and policies for pedagogical integration of social media and development of the best methods for teaching and learning based on these strategies.

KEY WORDS: *social media, higher education, university, scholar, faculty members*

I. The Social Media Landscape in Higher Education Context

Social Media is a generic broad term covering a large range of online platforms and applications which allow users to communicate, collaborate, interact, and share data (Doyle 2010; Zeng, Hall, and Pitts 2011) . It encompasses easily accessible web instruments that individuals can use in order to talk about, participate in, create, recommend, and take advantage of information, in addition to providing online reactions to everything that is happening around them.

Given the dynamic nature and complexity of social media it becomes quite difficult to define the concept. According to (Kaplan and Haenlein 2010) the confusion is even bigger among educational managers and academic researchers. Even we are not sure what is anymore (Malita 2011), we consider social media as today's most transparent, engaging, and interactive shift in education, "a group of Internet-based applications that build on the ideological and technological foundations of Web 2.0, and that allow the creation and exchange of user generated content" (Campbell 2010). Thus, social media is about transforming monologue into dialogue, about free access to all types of information, about transforming Internet users from mere readers to creators of content, about interacting in the online world so as to form new personal or business relationships.

Often used interchangeably with Web 2.0 we encounter social media on many different forms (Doyle 2010) like blogs, microblogs, social networks, media sharing sites, social bookmarking, wikis, social aggregation, virtual worlds, social games, and so many other (social) online artifacts. Nevertheless social media remain the communication and collaboration media that have registered the most important growth during the past year.

With the emergence/increased use of social media tools, a large number of higher education institutions are embracing this new ecology of information (Campbell 2010). More and more colleges and universities from all over the world are transitioning from traditional learning toward learning 2.0, widening their curriculum landscape beyond technology by integrating different forms of social media (Grosseck and Holotescu 2011b). Although in the literature there is no specific educational oriented definition, Conole and Alevizou (2010) give an indication that in order for learning 2.0 to occur, it is necessary to rethink the social academic work environments based on social media tools, in accordance with the learning needs, skills, and competencies of students (Wheeler 2010; Schaeffert and Ebner 2010).

The authors believe that it is important to get to know the specific characteristics of the audience of these social platforms, the applications and tools provided, with the aim of drawing correct usage and promotion principles that are applicable in the academic environment. Thus, the following section will discuss the findings of a mini-research undertaken by the authors within a broader project concerning the role of social media in the Romanian higher education context.

II. Research Methodology

2.1. Objectives and questions
The purpose of this mini-study is to gather information on ways in which academic staff are adopting social media platforms and to identify best uses. To ensure this objective is met, the following research questions are proposed: *How faculty members use social media as reflective and collaborative teaching and learning tools, also for research and professional development? Which are the potential benefits, challenges, and disadvantages in using social media in universities? How the usage can be extended, is there a need for training the educational actors in this topic?*

2.2. Method
For collecting the necessary information, we conducted an online questionnaire, publicized via academic networks of the authors' universities, relevant academic mailing lists, personal learning networks, as well as Twitter and Cirip, Facebook, LinkedIn, and other social web platforms.

Data collecting was performed between the end of February and the beginning of March 2012, with 79 respondents/answers, after validation. Because only a few people from our networks re-sent the link to the questionnaire, it was difficult to calculate the response rate.

III. Summary of the Findings

3.1. Respondents profile

Based on the findings obtained from the sample group we will begin with basic information about respondents' profile. *Who are they?* By gender 41 are male (52%) and 38 female (48%). By age the higher percent is allocated to the population between 36 and 45 years old (37%), 43% having less than 35 years.

What is their role in higher education? We managed to attract a wide variety of respondents at different stages of their academic careers: Professor—5% (4); Reader—15% (12); Senior lecturer—19% (15); Junior lecturer—14% (11); Researcher 5%—(4); Professor doctorate coordinator—1% (1); Academic administrator/Faculty development—4% (3); Other—36% (29). Where "Other" includes respondents who are in non-academic positions such as librarians, admission officers, trainers/instructors, doctoral candidates, or master students, etc.

What is their academic profile? While at first glance the results suggest that the categories were not comprehensive enough, we tried to cover all disciplines ranging from mathematics to medical sciences. Thus, almost half of the respondents (43%) aligned themselves with the exact science disciplines (i.e. mathematics, physics, biology, informatics, engineering, and earth sciences). Twenty-four percent (19) identify themselves as aligned with a discipline of social sciences (psychology, education, social work, political sciences), 13% are related with medical domain, 8 persons are humanistic oriented (foreign languages, philosophy, journalism, law), and only 8% are in the economic area (management, marketing, human resources, public relations, administrative issues, etc.).

We did not take into consideration some demographic characteristics such as: how many years a member of staff worked in higher education, the type of institution (college/university, public or private), size of the organization, tuition /without fees, etc. — these issues will be addressed and detailed in a future research.

3.2. Social media accounts profile

A second group of questions collected data about the specific social media platforms on which the responders are active, how they use them and what are the benefits and limits encountered. On most social media platforms 90% of users are passive lurkers who never contribute, 9% are active lurkers who reshare or comment, while only 1% are content creators or co-creators (Nielsen 2006). *Do Romanian educational actors follow this Social Media Engagement Rule?*

The question *"How do you use the following social media?"* refers to the use only for documentation or also for content creation of a large area of networks and social media platforms. The analysis of these large categories, covering the current social media landscape (Solis and JESS3 2008), makes an important difference between our investigation and other recent studies (Faculty Focus 2011; Moran, Seaman, and Tinti-Kane 2011).

SOCIAL MEDIA USAGE			
Social media networks and applications around content used for	**Document-ation %**	**Post notes/content %**	**Not a user %**
Blog (any type of platform/Blogger, WordPress, weblog.ro)	22	44	34
Miniblog (Tumblr.com, Posterous.com)	14	6	80
Microblog (Twitter.com, Cirip.ro, Plurk.com, Edmodo.com)	19	29	52
General Social Networks (Facebook.com, Plus.Google.com, MySpace.com)	10	68	22
Professional Social Networks (LinkedIn.com, Xing.com, Academia.edu)	28	48	24
Social Bookmarking (Delicious.com, Diigo.com)	10	23	67
Video sharing (Youtube.com, Vimeo.com, TED.com, TeacherTube.com, Trilulilu.ro, MyVideo.ro)	46	43	11
Image sharing (Flickr.com, Picasa.Google.com, deviantART.com)	29	49	22
Audio/Podcasting sharing (Blip.fm, SoundCloud.com)	10	10	80
Presentation sharing (Slideshare.net, Authorstream.com, Prezi.com)	22	39	39
Document/Books sharing (Scribd.com, DocStoc.com, Docs.Google.com, Books.Google.com)	32	56	13
Mindmaps (Mindomo.com, Mindmeister.com, Spicynodes.org)	6	18	76
Screencasting (Screenr.com, ScreenJelly.com, ScreenCastle.com)	4	13	84
Livestreaming (Qik.com, UStream.com)	6	9	85
Feeds Monitoring (Reader.Google.com, Bloglines.com)	24	24	52
Wiki (Wikispaces.com, MediaWiki.org, Wikia.com, PBWorks.com)	44	34	22
Digital storytelling (Voicethread.com, Glogster.com, Capzles.com, Notaland.com, Storybird.com, Storify.com, Photopeach.com, Projeqt.com)	0	15	85

Almost all of the respondents are aware of the large categories of platforms. The most popular seem to be those for multimedia content sharing: video—89% of responders declared that they use such platforms, documents/books —87%, image—78%, in all cases at least half posting content. The large interest for the documents/books sharing (78%) and presentation sharing platforms (61%) confirms the social reading trend in the 2012 Horizon Report in higher education. However, we can note that the platforms for podcasting and audio sharing are at the opposite pole of interest —only 20% of the respondents use them.

More than two-thirds are active on wikis (78%), general networks (78%), professional networks (76%) and blogs (66%), and more than half of them post content on these platforms, the highest rate of postings being on general networks (68%). Half of the respondents (48%) monitor feeds to keep track of news and activate on microblogs. As one of the most important uses of microblogging is for news searching (56% in (Grosseck and Holotescu 2011a)), the micro-posts streams can be seen as curated feeds, containing news, but also comments and validation. Only 20% pay attention to miniblogs (such as Tumblr and Posterous). Even if with very interesting and challenging uses, such as collaborative work on scenarios, tutorials, and micro-lectures, the educators show low interest in mindmapping (24%), screencasting (16%), or digital storytelling platforms (15%). An explanation could be the fact that to use such platforms you need to be and stay informed, to activate in online communities where one needs to learn and share ideas and experiences.

Calculating an average for all of the platforms, we can affirm that 31% of the respondents create content, a percentage much higher than that of 9% for active lurkers and 1% for creators. But before concluding that the Romanian educational actors are breakers of the "Social Media Engagement Rule" (Nielsen 2006), we should not forget that the questionnaire responses were received from active users who wanted to get involved in this research approach.

PLATFORMS FOR COMMUNICATION/COLLABORATION/LOCALIZATION		
Do you use the following social media for communication/collaboration/localization?	No	%
Groups (Groups.Google.com, Groups.Yahoo.com, Ning.com, Meetup.com)	71	90
Forums/Spaces for discussions(phpBB.net, Quora.com, Disqus.com)	26	33
Localization (Foursquare.com, Yelp.com, Zvents.com)	8	10
Augmented reality (Layar.com, Wikitude.com, Zooburst.com)	6	8
Virtual worlds/Social Games (Secondlife.com, Playdom.com, OpenSimulator.org)	7	9
IM (YM, GTalk, Jabber, Skype)	53	67

If the groups or IM tools, which can be considered as Web 1.5 applications, are used by a large majority (90%, respectively 67%), the new discussions applications, such as Quora or Disques, are known to only 33% of the respondents, localization for 10%, augmented reality (AR) for 8%, and virtual worlds/social games for 9%. These figures can be correlated with the issue that the experience in integrating such tools in education is lower, also with the fact that the applications for localization and AR are mobile, and we will see that a relatively low percentage of educators use mobiles or tablets/ipads.

At the question *"What other social media tools/categories do you use?"* even if only a few answers were received, they are very interesting and worth mentioning: collaborative graphs and infographs, desktop sharing applications (BeamYourScreen), eLearning platforms (Moodle, Sharepoint) with social media features, platforms for academic research (Researchgate), for social learning (Schoology), for project management (Basecamp), or for software engineering (GitHub).

ARE THE FOLLOWING STATEMENTS TRUE FOR YOU ?			
Statements related to social media	Yes (%)	Not yet, but I am aware of it (%)	No (%)
I access social media via mobile	46	27	28
I access social media via tablet/ipad	15	48	37
I evaluate the activity of my students on social media platforms	30	27	43
My institution assesses my activity on social media platforms	15	24	61
My institution encourages/supports the usage of social media by teachers/students/pupils	34	30	35
My institution has specific policies related to social media usage	15	37	48
I became familiar with SM during a course/workshop/project	30	4	66

Almost half of the respondents access social media platforms using mobile phones, while only 15% are equipped with tablets/iPads. A third (28%, respectively 37%) seem not to be interested in using mobiles or tablets/iPads for this purpose.

The percentage of teachers (30%) who evaluate the activity of their students on social media platforms is very close to that of teachers (34%) coming from institutions which encourage and support the use of social media by teachers/students/pupils. However, we can note that the institutions of only 15% of responders assess their activity on social media platforms or have specific policies related to social media usage. Even if only one-third of educational actors became familiar with social media during a course, workshop, or project, a very low percentage (4%) are interested in participating in such a training. A breakdown of educational actors awareness in using social media in different activities appears in the following table.

DO YOU USE SOCIAL MEDIA FOR THE FOLLOWING ACTIVITIES?			
Activities	Yes—I have used	Not yet, but I am aware of it	No
Didactic activities	61%	18%	22%
Research activities	58%	20%	22%
Professional development	**78%**	**11%**	**10%**
Personal development	**78%**	**8%**	**14%**

The greatest percentage (78%) is using social media for professional and personal development, while high percentages are also for those who use such tools for didactic activities (61%) and research activities (58%). We can say that there is a true adoption of social media in all the domains of the educational process, the rate being much higher than that concerning only the specific technology of microblogging (Grosseck and Holotescu 2011a). The survey showed that there is a relatively small group of educators (10–22%) who believe that social media has no place in education.

Regarding the mode of communication and collaboration we see that social media are a medium used at all levels, with peers from their own country or abroad, by around two -third of responders. Again the percentages are much higher than those for microblogging, which still has a narrow adoption (Grosseck and Holotescu 2011a), the same note is available for the next question too. What seems surprising here is that the lower level of own department/faculty (with the highest f2f interaction) is the one where social media tools are highly used, by 77% of responders.

LEVELS OF COMMUNICATION/COLLABORATION		
I work with ...	Number	Percent
Peers from different institutions from Romania	52	66
Collaborators in different institutions from other countries	47	59
Colleagues/peers across my university/institution	49	62
Peers and Doctoral and Master students of my own department/faculty	61	77

The following table includes what our study has revealed regarding the most common types of uses of social media by the scholarly community.

CONTEXTUAL CONDITIONS IN WHICH SCHOLARS USE SOCIAL MEDIA		
Activities	Number	Percent
Searching news, academic content	70	89
Dissemination of own results, articles, projects, presentations	49	62
Inquiring/research (reviewing literature, collecting/analyzing research data)	52	66
Personal/Professional Communication/Collaboration	65	82
Networking for professional development	36	46
Building a community of practice	24	30
Building a learning community with students enrolled in formal courses	30	38
Participating/following different scientific events (as a real time news-source)	52	66

The findings indicate that social media usages by educational actors are:

- *Search for scholarly content*—the highest percentage of responders (89%) is looking to discover news, ideas, experiences, articles, and projects.

- *Dissemination channels* for promoting own results/articles/projects or presentations —appreciated as being powerful by 62% of the respondents.

- 66% say that social media tools are important in *reviewing the literature, collecting, and analyzing research data.*

- *Sharing professional experiences online*, communicating scholarly ideas, collaborating with peers or with networks of stakeholders are favorite activities for 82% of users.

- Building a *network of contacts* for research opportunities, for finding sponsors or for reaching fellow specialists was indicated by 46% of the responders.

- Less than one - third (30%) appreciate the power of sharing, skills development, or knowledge creation by building *communities of practice*.

- A percentage of 38% show a low interest in building *learning communities*, student centered. Thus we can say faculty members are (still) unprepared to integrate social media in their courses.

- Nowadays, *following* presentations, livestreamings, videos, and posting from *scientific events* is a common practice, adopted by two-third of responders (66%).

The questionnaire has also two open-ended questions asking respondents to list/to identify main advantages and constraints to uptake when using social media in higher education. Almost all of the respondents share their impressions, which ranged from positive general comments to negative remarks, like "I think social media are very useful for communication and collaboration" to "I just don't get it".

Although social media redefine the relationship between technology and education, using them in academic courses does not represent an easy teaching/training/researching and learning method. It implies a sum of efforts, and especially knowledge of these technologies, with both benefits and limits.

Advantages expressed by participants:
- *accessibility and ease of use* (anyone can create a blog or a YouTube

account in just a few minutes), including mobile devices and applications (smartphone, tablets, qr-codes, augmented reality, etc.);

- *cost reduction* (low educational marketing costs)—most social media sites offer access to services, information and community free of charge;
- *flexibility, transparency and autonomy of applications*;
- *educational "recruit ability"* in social networks (the results support what (Barnes and Lescault 2011) study documented: higher education institutions are e s p e c i a l l y using social networking sites, not only to recruit but to research prospective students);
- *changing teachers' attitudes* toward using social media in academic courses (taking academics out of their usual comfort zone);
- *engaging/enriching/empowering students' interactions and participation* through the use of social media in academic environments;
- *collaborative characteristics/features* which erase the barriers between formal and in/non-formal learning;
- *establishing relationships and conversations* among teachers, students, professionals, and researchers from different institutions;
- *facilitating learning* through personal learning networks / environments (peer-to-peer learning and mentoring);
- *social interactions* in communities for learning, practicing, as well as professional ones (learning from experts and peers);
- *teaching / learning digital skills* like creation, curation, and sharing online/digital content/knowledge;
- *easily-accessible creativity*/accumulative information;
- "*use of authentic study materials*", some of them in real-time (i.e. microblogging is an easy way to engage in dialogues with anyone, for instance);
- a non-conformist and flexible academic environment ("*easy socialization*");
- facilitating the processes of providing information, of building knowledge ("*a modern approach of educational subjects*");
- *feedback* (one can receive ideas, suggestions, and opinions from mere visitors, one can update the strategy or educational services, or improve the course);
- *easy monitoring online presence and reputation*;
- *collaborative participation*—developing research projects at a distance;
- using *open education* in terms of: open source/free software, open educational resources, open content, open access publication, open teaching, a n d open scholarship.

Almost all of the respondents highlighted barriers or limits of using social media in higher education. Based on their responses, it appears that most of the comments are related to the following *disadvantages*:

- *content trivialization* caused by a lack of validation procedures (the crowdsourcing effect);
- *security of data and persons; aggressive/mistrusted/unfiltered*

information flows (one of the respondents said :"*it has the same taste as an unfiltered beer*");

- *online information / cognitive overload,* advertising interference, informational abuse, spam, disorientation, infoxication, fragmentation, etc.;

- *equality or e-quality* (anyone can publish web content, but not everyone offers quality content; unsolicited content); *neglecting the educational goals* / purposes / social limitations;

- *difficult management of digital identity/anonymity*: fake IDs and hiding one's real identity have been and will continue to be issues;

- *ethical concerns:* proper professional behavior in the use of social media: confidentiality, defamation, following university regulations/the academic social media policy;

- *institutional norms*/terms of use and best practices in the field, disadvantageous policies for educational sector (i.e. in Romania there are no academic clear rules regarding the use of social web tools in education; there is also a need to have a unique platform for the entire university/professional staff);

- *time spent on social media sites*: all things require time and dedication, and social media entails online presence, dialogue, and sustained activity;

- social media also requires *a certain life style and/or an organizational culture* in the digital era; *emotional barriers*: perceptions of technology, anxiety related to its use, lack of confidence in their potential, and negative personal experiences related to technology

- *artificial communication*: written communication versus oral communication (f2f versus online);

- *the noise:* pseudo-relationships, in-appropriate reactions, personal exposure, etc.;

- *the activity with/within social media is not recognized as academic* (more specifically —it does not count in periodic assessment).

For the time being, we can say that only a few universities have adopted coherent strategies for pedagogical integration of social web functions and development of the best methods for teaching and learning based on these. Thus, for a more accurate picture of social media landscape in academia it is necessary to repeat the study at least for several years to provide a longitudinal look at adoption of social media by colleges and universities.

It is also necessary to build online communities for professional learning, academic practice, quality, and leadership for managers of institutions, as well as for the people involved in both teaching and administration. There should be more social media platforms dedicated to communities of education experts (policies, foresight, etc.), there should be an institution-wide *Social Media Observer* that strengthens university policies related to social media at the level of the higher education institution and that represents, at the same time, a landmark for strategic positioning of universities within the new technological landscape. However, an informal social media educational platform, functioning in conjunction with the official platform, will not only become an extremely efficient communication channel, but will also emphasize the culture of the students and that of the staff of the institution in question. The most important type of feedback will continue to be interactivity.

IV. Conclusions

Despite social media popularity among staff (Merrill 2011) and its predominantly positive perceptions among higher education institutions, the use of social media "does not come easily" (Harris and Rea 2009) and is still at the level of experimentation, as it is trying to find its place in the online environment of Romanian higher education area. In the meantime, academia must free itself from its fears, prejudices, and arrogance. In order for this to happen, the management of higher education institutions must change, firstly by acknowledging the need to have a social media presence, and then by providing clear regulations regarding its use (private life, protecting intellectual property, etc.). It is also important to recognize the importance of social media in the recruitment of students, dissemination of research, and brand building (alumni included), as an engagement tool and not as a megaphone (Colvin 2011). Furthermore, we need assigning social media responsibilities within faculties and departments. Thus, the organizational charts of our institutions should include "new" positions such as: learning architect, learning/social media community manager, serious game designer or learning autonomy counselor (Grosseck and Holotescu 2011b). Perhaps the most significant approach of using social media in universities is the fact that it is more a socio-cultural phenomenon, rather than a technical one, an attitude rather than a sum of technologies, the fact that it has become more personal to the students, emphasizing the development of communities of learning and practice and the strength of something done together.

To conclude: We believe it is necessary that *a social media education* be accompanied by *social media in education*.

About the Authors
Carmen Holotescu is the Director of Timsoft eLearning, Romania
Gabriela Grosseck ia an Associate Professor at University of the West, Timisoara

References
Allen, J.P., H. Rosenbaum, and P. Shachaf. 2007. "Web 2.0: A Social Informatics View." In Proceedings of the Thirteen Americas Conference on Information Systems, Keystone. https://usffiles.usfca. edu/FacStaff/jpallen/public/we2_si-draft.pdf.
Barnes, N.G., and A.M. Lescault. 2011. Social Media Adoption Soars as Higher-Ed Experiments and Reevaluates Its Use of New Communications Tools, Center for Marketing Research, University of Massachusetts Dartmouth, North Dartmouth, MA.
       http://www.umassd.edu/cmr/studiesandresearch/socialmediaadoptionsoars/.
Bradwell, P. 2009. The Edgeless University. Why Higher Education must Embrace Technology. UK: Demos.
Campbell, D. 2010. "The New Ecology of Information: How the Social Media Revolution Challenges the University." Environment and Planning EPD: Society and Space 28(2): 193–201. http://www.envplan.com/abstract.cgi?id=d2802ed.

Colvin, S. 2011. *What is the State of Social Media in Higher Education?*, blog post. http://uv- net.uio.no/wpmu/hedda/2011/07/15/what-is-the-state-of-social-media-in-higher-education/.

Conole, G., and P. Alevizou. 2010. *A L iterature Review of the Use of Web 2.0 Tools in Higher Education.* A report commissioned by the Higher Education Academy, The Open University Walton Hall, Milton Keynes UK. http://www.heacademy.ac.uk/assets/EvidenceNet/Conole_Alevizou_2010.pdf.

Doyle, C. 2010. *A Literature Review on the Topic of Social Media.* http://cathaldoyle.com/wp-content/uploads/2010/10/Literature-Review-of-Social-Media.doc.

Faculty Focus. September 2011. *Social Media Usage Trends Among Higher Education Faculty.* Special Report. A Magna Publication. http://www.facultyfocus.com/wp-content/uploads/images/2011-social-media-report.pdf.

Freire, J., and K.S. Brunet. 2010. "Políticas y prácticas para la construcción de una Universidad Digital." *La Cuestión Universitaria* 6: 85-94. http://www.lacuestionuniversitaria.upm.es/web/grafica/articulos/ imgs_boletin_6/pdfs/LCU-6-7.pdf.x

Grosseck, G., and C. Holotescu. 2011a. Academic research in 140 character or less. Paper Presented at *7th International Scientific Conference "eLearning and Software for Education"*, Bucharest, April 28–29, 2011. http://adlunap.ro/eLSE_publications/papers/2011/1590_1.pdf.

Grosseck, G., and Holotescu, C. 2011b. "Social media challenges for Academia." In *Contemporary Issues in Education and Social Communication. Challenges for Education, Social Work and Organizational Communication*, eds. B. Pǎtruţ, L. Mâţǎ, and I.L. Popa. München: AVM—Akademische Verlagsgemeinschaft München, 148-174.

Harris, A., and A. Rea. 2009. "Web 2.0 and Virtual World Technologies: A Growing Impact on IS Education." *Journal of Information Systems Education* 20 (2): 137-144.

Hartman, J.L., C. Dziuban, and J. Brophy-Ellison. 2007. Faculty 2.0. *EDUCAUSE Review* 42(5, September/October): 62-77.

Kaplan, A.M., and M. Haenlein. 2010. "Users of the World, Unite! The Challenges and Opportunities of Social Media." In *Business Horizons* 53(1): 59-68.

Lee, M.J.W., and C. McLoughlin, eds. 2011. *Web 2.0-Based E-Learning: Applying Social Informatics for Tertiary Teaching.* Hershey, New York (Information Science Reference)

Liu, M., D. Kalk, L. Kinney, G. Orr, and M. Reid. 2009. "Web 2.0 and Its Use in Higher Education: A Review of Literature." In *Proceedings of World Conference on E-Learning in Corporate, Government, Healthcare, and Higher Education 2009*, eds. T. Bastiaens et al., Chesapeake, VA: AACE. 2871-2880. http://www.editlib.org/p/32892.

Malita, L. 2011. "Social Media Time Management Tools and Tips." *Procedia Computer Science* 3: 747-753.

Merrill, N. 2011. "Social Media for Social Research: Applications for Higher Education Communications." In *Higher Education Administration with Social Media (Cutting-edge Technologies in Higher Education, Volume 2)*, eds. Laura A. Wankel and Charles Wankel. Emerald Group Publishing Limited, 25-48.

Moran, M., J. Seaman, and H. Tinti-Kane. 2011. *Teaching, Learning and Sharing. How Today's Higher Education Faculty Use Social Media.* Pearson Learning Solution and Babson Survey Research Group. http://www.pearsonlearningsolutions.com/educators/pearson-social-media-survey-2011-bw.pdf.

Nielsen, J. 2006. *Participation Inequality: Encouraging More Users to Contribute.* http://www.useit.com/alertbox/participation_inequality.html.

Schaeffert, S., and Ebner, M. 2010. New Forms of and Tools for Cooperative Learning with Social Software in Higher Education. *In Computer-Assisted Teaching: New Developments*, eds. B.A. Morris, and G.M. Ferguson. Nova Publishing, 151-156.

Solis, B., and JESS3. 2008–present. *The Conversion Prism.* http://www.theconversationprism.com/.

Wheeler, S. 2010. "Digital Tribes and the Social Web: How Web 2.0 Will Transform Learning in Higher Education." *Keynote Speech, Engaging the Digital Generation in Academic Literacy: Learning and Teaching Conference*, University of Middlesex, England. 29 June, slides. http://www.slideshare.net/timbuckteeth/digital-tribes-and-the-social-web.

Zeng, L., H. Hall, and M.J. Pitts. 2011. "Cultivating a Community of Learners. The Potential Challenges of Social Media in Higher Education." In *Social Media: Usage and Impact*, eds. H. Noor Al-Deen, and J.A. Hendricks. Lexington Books.

Interoperability Strategies For Serious Games Development

Ioana Andreea Stănescu
"Carol I" National Defence University, Romania

Antoniu ªTefan
Advanced Technology Systems, Romania

Milos Kravcik
RWTH Aachen University, Germany

Theo Lim
Heriot-Watt University, Scotland

Rafael Bidarra
Delft University of Technology, Netherlands

Abstract*: Serious games have emerged as a new medium that enables players to acquire and enhance their skills and knowledge particularly in education and increasingly across a spectrum of fields from industrial and emergency training to marketing. While the use of serious games has extended rapidly to a variety of domains, their design and development remains a challenging process both for developers and teachers/trainers. This paper approaches the technological environment underpinning the development of serious games, and focuses on interoperability as a core element of a sustainable endeavour. Developing serious games in a way that enables interoperability is one means of increasing the depth and scope of instructional materials available to learners while reducing the overall development costs and time. Interoperability, the ability of computers and applications to communicate and share resources in a heterogeneous environment, is dependent on standards. Optimizing requirements of accessibility, interoperability, durability, and reusability for maximizing cost efficiency start with a proper understanding and integration of standards. The authors argue that interoperability provides a context for the development of sharable education resources and technologies which in turn allow for collaborative education in a field in which rapid technological developments are making it difficult for instructors and developers to stay up-to-date with both the science and the related technologies. The paper analyses various serious games interoperability scenarios and address the main gaps surrounding standardization in this field with the purpose of assisting developers and teachers in implementing successful solutions. The scenarios are based on a Serious Game Multidimensional Interoperability Framework that integrates three key dimensions: the core components included within a serious game (game mechanics, gameplay, graphics engine, and graphic objects), the ecosystem where the serious game will be implemented (developing platforms, programming languages, and LMS communications), and external factors that go beyond the core technical aspects of a serious game (assessment, applicability, classification, and glossary of terms). The research considers the existing standards—such as SCORM and LOM—that impact serious games development, as well as gaps and fragmentation issues that hinder the development process with the purpose of identifying efficient, adaptable solutions.*

KEY WORDS: *serious games, interoperability, SCORM, LOM, LMS, SG-MIF*

I. Introduction

Interoperability is one of the core themes of serious game (SG) development (Stănescu, ªtefan, and Roceanu 2010; Bergeron 2006) and it aims to support an effective exchange of information based on consistent, specific data and technical standards. Interoperability scenarios aim to enhance the interaction between serious game developers by means of alternative technological solutions that are derived from the standards. Paraphrasing the famous quote of George E. P. Box, *all models are wrong, but some are useful,* and considering the fact that serious game developers reside not only in academic, but also in industry environments, it can be concluded that no standard or scenario for interoperability can constitute the ultimate solution or the panacea for serious game development. Therefore, this paper analyses various challenges and different interoperability scenarios that coexist within serious games ecosystems with the purpose of enabling adaptive solutions. This research aims to facilitate in-depth understanding of cost-efficient development and large-scale implementation of reusability in serious games (SG) environments based on a *Serious Game Multidimensional Interoperability Framework (SG-MIF).*

1.1. Identifying the need for interoperability

Interoperability is a key requirement for organizations regardless of the field they operate in—education (Spires 2008), commerce (Panetto, Scannapieco, and Zelm 2004), health care (Benson 2009), or military (Roberts and Gallagher 2010). Many educational organizations already operate large environments that implement different technical solutions. When these organizations perceive the need for new/additional applications implemented within their environments, the automatic tendency is to start thinking based on currently implemented environments. This is usually referred to as Technology Aligned Environment, where decisions about enhancing the current environment are more closely connected to what is already running rather than on the basis of which provides the best platform (Microsoft Corporation 2004).

With the ever-increasing requirements for efficiency, responsiveness, and cost reduction, interoperability stands as a core demand for modern IT environments. The European Interoperability Framework stands out as an effort to facilitate the delivery of eGovernment services to citizens and enterprises within a multi-vendor, multi-network, and multi-service European area. It has emerged from the necessity to support the development of the single market where European public administrations are interoperable to enable any supporting information exchanges (Commission of the European Communities; Interoperability Solutions for European Public Administrations).

A sustainable, flexible development of serious games employs vehicles that enable the ability of serious game components and of serious games ecosystems to work together easily and effectively by design. The research on serious games interoperability focuses on five key areas:

- *Standardization.* Analyze scenarios that enable the creation of functionally interchangeable items, while considering opportunities, challenges the existing standards, and best practices that impact SG development.

- *Interchangeability.* Identify methods that would make game components interchangeable, without having to alter the item to make the new combination possible.

- *Standards adoption.* Analyze the position of the development companies and of the educational actors toward standards adoptions in an effort to create adaptable solutions.

- Open systems architecture. Provide a modular design that defines key interfaces within a system using widely supported, consensus-based standards that are available for use by all developers and users without any proprietary constraints.

- Unique specifications and proprietary devices. Consider the fact that unique or utility- specific applications and vendor-proprietary applications and devices can be counter-productive to interoperability, but may be necessary to provide needed functionality.

1.2. The whys of serious games interoperability failures

The problem of incompatibility due to multiple hardware platforms, operating systems, and languages impacts upon the serious games development environment. At the moment, there are thousands of simulations, teaching programs, and also games that cannot interoperate (González 2011). Such systems need to be highly interoperable, easily configurable, aligned, and consistent with local and global efforts (Roman a n d Bassarab 2008). The experience of educational communities illustrates not only the need for standards, but also the need for adaptable interoperability scenarios.

Common interoperability standards could benefit both the academic and developer communities, enabling them to solve common problems with common solutions. At present, there is no consensus in the games industry on the desirability of a common set of interoperability standards (Bergeron 2006). Resistance to common interoperability standards is generally based on the following factors:

- Technical considerations: Common standards accommodate a wide range of potential users and therefore are not optimal for any particular use. Many game companies prefer to design custom protocols that maximize performance.

- Not-invented-here syndrome: many commercial firms have a bias against technology developed outside their own organization.

- Strategic value of proprietary solution: proprietary protocols are viewed as a strategic competitive advantage. Use of a public standard would eliminate one element of advantage by allowing competitors to use the same technology. In addition, use of a public standard could signal that a company is unable to develop a better solution.

- Control: Adoption of an industry or public standard reduces the control a company has over its protocols. Standard committees determine changes to the protocol. Companies that control their own protocols can upgrade them at their own pace, as the need arises.

Even if game developers are willing to examine protocols for suitability in their games, few have actually implemented them. Some companies find protocols too big and complex, performing operations that were not relevant to games and slowing the performance of the system. Others prefer to develop derived protocols that include only those functions needed to support their applications.

Each of these implementations is proprietary to the developing company and not interoperable with other companies' protocols.

Besides the direct technical considerations, standard and interoperability failures relate to collateral elements that impact upon their success. Decisions in this area are made by private, standard bodies, and industry consortia that operate largely outside of the public eye and with little input from public interest groups or public policymakers (Morris 2 0 1 1). For information and communication technology standards resulting from these private processes to meet any comprehensive definition of "openness", standard developers need to consider and reflect the input from public policy experts.

Moreover, history has shown interoperability to be also a people problem, the people's failure to use best practices to develop the right processes and tools (including standards) for sharing trusted information (Desourdis, Rosamilia, and Jacobson 2009).

II. Interoperability in Serious Games Ecosystems

Previous researches have focused mainly on interoperability issues of singular components, like game engines (Ryan, Hill, and McGrath 2005; Stǎnescu et al. 2011), while specific technical areas, such as distributed simulations like HLA have not been taken into consideration. This research takes a holistic approach and builds upon three key elements that impact upon serious games interoperability: the components included within a serious game, the ecosystem where the serious game will be implemented and external factors that go beyond the core technical aspects of a serious game (Ryan, Hill, and McGrath 2005). These elements form the core research dimensions of a Serious Games Multidimensional Interoperability Framework (SG-MIF). The researchers consider that such frameworks enable prior evaluation of alternative interoperability scenarios by providing an overview on interoperability-based SG development. The following sections detail upon different interoperability scenarios extracted based on the SG-MIF.

2.1. Serious games and standards: SCORM and LOM

SCORM (Sharable Content Object Reference Model) is a standard developed by ADL (Advanced Distributed Learning) that enables sharing of distributed learning content across SCORM compliant learning management systems. The main questions arising in connection with the SG–SCORM relationship are: What types of information can be exchanged using SCORM? What types of serious games components can be reused using SCORM?

The SCORM specification covers two particular topics related to serious games: package and deployment, and communication between serious games and Learning Management Systems. In this way, an SG is conceived as a SCO object, and considering the SCORM Content Aggregation Model it can be deployed in multiple commercial and open source LMS platforms already available. In addition, SGs can generate a great amount of tracking information that can be used by the instructor to evaluate the student play session. Using the SCORM Runtime Model an SG can set some of the cmi.* properties: cmi.completion_status; cmi.success_status; cmi.core_score_raw; and cmi.interactions. cmi.interactions is a collection of properties, that is, multiple values can be collected inside this property. In contrast to cmi_score_raw and cmi.success_status that provide a coarse-grained evaluation of the student's performance, cmi.interactions.* can be used to provide a fine-grained or detailed report of the student game play session and its relation to the SG learning objectives.

These properties (and the rest of the SCORM data model) can be used in game engines such as the e-Adventure authoring tool. This way, the internal game state can be translated to a platform neutral data model. Moreover, e-Adventure games sent the information back to the LMS using the SCORM Runtime API, so the game tracking information can be reviewed or used by other tools that are hosted in the LMS.

The IEEE Learning Object Metadata is a standard metadata schema that aims to provide a common vocabulary to describe e-learning content materials. In relation to SG, two key questions arise: How can serious games employ the standards defined by LOM to enable learning content classification? Is the use of metadata a feasible solution for serious games? If serious Games are to be considered as a particular case of learning objects, fostering the reuse of existing SG by adding metadata to them seems a logical path to follow. Taking into account the cost constraints related to SG development makes adding metadata to SGs a necessity that fundament reusability in SG environments (Ryan, Hill, and McGrath 2005).

2.2. Interoperability between games and Learning Management Systems

The key role of a modern Learning Management System (LMS) is to facilitate the interaction between tutors and learners, detailed tracking of the students' progress, and a simple path for the delivery of content through the web. Therefore, it is important to enable the exchange of information between serious games and the learning management system(s), with the purpose of tracking user progress and behavior. Given the variety of game engines, programming languages, and hardware platforms, how can interoperability between this variety of SGs and LMSs be enabled?

The game state can be changed by stimulus, such as a mouse, gamepad, or game dynamics. Every game must have a container for game objects. There are two kinds of objects: one belongs to the game logic and is called an actor and the other belongs to the renderer and is called a textured skeletal mesh. If the game state of the actor changes, the game logic sends an event to the renderer and it reacts to this event by changing the texture. To conclude, the game logic holds the object state and the game view holds model data and textures. SGs have more than just storyline, design, and software. Pedagogy in this type of game plays a major role. For this reason we need a new component to check and report when the learning objectives have been met by the learners. The added component is called game tracking layer.

There is little research related to the interoperability between SGs and LMS platforms, most of them use the SCORM specification to package and deploy web-based SGs and to send and receive information from the LMS using the SCORM API. There is no standardized specification or standard to integrate desktop games with an LMS platform; however, there is a research trend related to the SCORM High Level Architecture, aiming to integrate training simulation software with an LMS.

2.3. Interoperability between game components

In the field of SGs, the interoperability between game components has the role to sustain the reusability of basic multimedia assets and game objects of low, medium, and high complexity. When referring to reusability of game components we can identify two main groups of components:

1. Basic multimedia assets that are reused and integrated into the game based on their file format, and the capabilities of the game engine. Examples of these assets: images, 3D objects, and audio and video clips. These multimedia assets have the highest reusability potential of all of the other game component types because they do not require complex prerequisites on the part of the game engine—the game engine just has to support the specific file format and they become available for use. Because these assets are part of a fixed class of types, with fixed properties, it is very easy to use conversion applications that transform one file format into a compatible one, without loss of functionality (e.g. an image can have the exact same properties regardless of its file format type).

2. Complex game objects that incorporate additional semantic metadata and even custom scripting code can be executed by the game engine interpreter to maximize customization capabilities, for example, a user avatar that defines properties such as

weight, gender, voice characteristics, strengths, and weaknesses. These classes of objects have specific prerequisites on the part of the game engine—besides the actual file format, the extended properties of these games must be described in a method that can be read by the game engine and interpreted according to their meaning. Which classes of game components can be made reusable? What is the best method of describing and embedding metadata about complex game objects so that they can be effortlessly integrated into a variety of game engines?

2.4. Serious Games Multidimensional Interoperability Framework

In the light of the above, it becomes necessary to advance a method that builds upon research carried within specific areas and that provides a clear overview of alternative interoperability solutions.

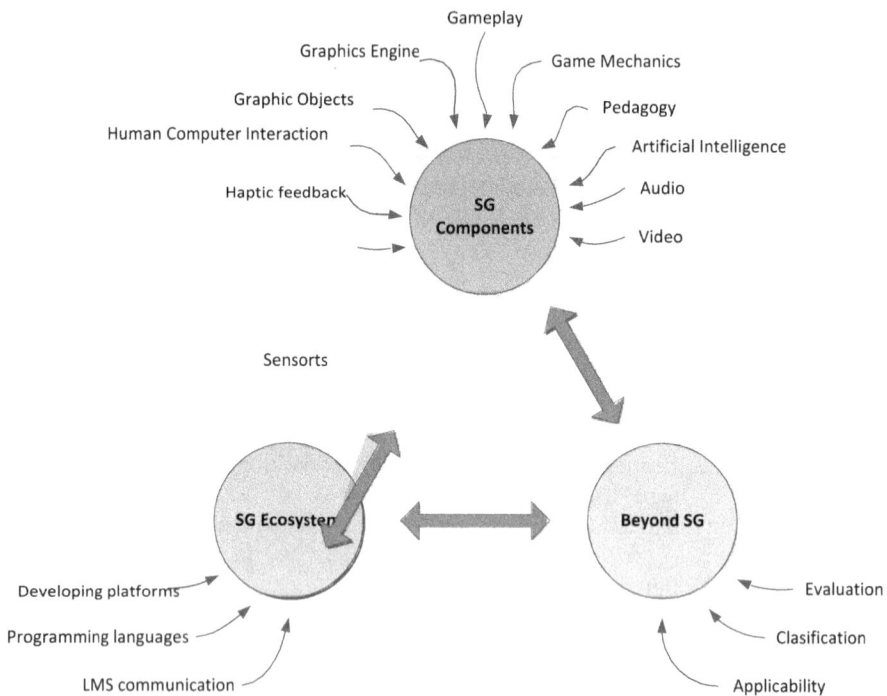

Figure 1. Extended Serious Games Multidimensional Interoperability Framework (SG-MIF)

This Serious Games Multidimensional Interoperability Framework (SG-MIF) has been developed to facilitate the in-depth analysis of this complex research topic, as well as to put into perspective the SG development ecosystem.

III. Conclusions

The approaches in the interoperability literature (Stănescu et al. 2011) do not propose mechanisms for prior evaluation of interoperability solutions. This paper advances an SG-MIF that aims to fundament prior evaluation, respectively, the ability to evaluate, earlier in a serious game project, two key elements: the potential improvement that will result from the solution's implementation; and the impact of this interoperability improvement on the achievement of the SG objectives at technical, operational, and strategic level. The researchers hypothesize that the difficulty of these approaches that aim to assess the contribution of interoperability to the SG development strategy relates to the fact that they do not take into account the entirety of the SG ecosystem. The goal of this research was to propose a framework that addresses this issue and that enables SG developers to consider different levels of interoperability, as well as develop adaptable interoperability scenarios. Future research will focus on mapping the strengths and the challenges of each of the sub-elements of the SG-MIF with the purpose of offering a clear picture on the advantages and disadvantages of each interoperability solution.

About the Authors

Ioana Andreea Stănescu is RDI Project Manager at Advanced Technology Systems in Romania.

Antoniu ªTefan works at Advanced Technology Systems in Romania.

Milos Kravcik is a research associate at Aachen University of Technology (RWTHG) in Germany.

Theo Lim is a lecturer in the School of Engineering & Physical Sciences at Heriot-Watt University, Scotland.

Rafael Bidarrais is an associate professor Game Technology in the Faculty of Electrical Engineering, Mathematics and Computer Science at of Delft University of Technology, The Netherlands.

References

Benson, T. 2009. *Principles of Health Interoperability HL7 and SNOMED (Health Informatics)*. London: Springer, 25.

Bergeron, B. 2006. *Developing Serious Games*. s.l.: Cengage Learning.

BinSubaih, A., S.C. Maddock, and D. Romano. 2007. "*A Survey of 'Game' Portability*," Tech. Rep. CS-07-05, University of Sheffeld, Sheffield, UK.

Camara M., Y. Ducq, and R.A. Dupas. 2012. "Methodology of Interoperability Evaluation in Supply Chains Based on Casual Performance Measurement Models." In *Enterprise Interoperability V: Shaping Enterprise Interoperability in the Future Internet (Proceedings of the I-ESA Conferences)*, eds. R. Poler, G. Doumeingts, B. Katzy, and R. Chalmeta. Springer, 3-15.

Commission of the European Communities. Linking up Europe: the Importance of Interoperability for eGovernment Services. http://ec.europa.eu/idabc/servlets/Doc2bb8.pdf?id=1675 (accessed on February 25, 2012).

González, A. 2011. *Interoperability System. Service Robotics within the Digital Home*, 53. s.l.: Netherlands: Springer, 1-47.

Desourdis, R., P. Rosamilia, and C. Jacobson. 2009. *Achieving Interoperability in Critical IT and Communication Systems*. Artech House Publishers.

Interoperability Solutions for European Public Administrations. Towards a Closer Alignment of Interoperability Frameworks across Europe. http://ec.europa.eu/isa/actions/04-accompanying-measures/4-2-3action_en.htm (February 25, 2012).

Microsoft Corporation. 2004. *Application Interoperability: Microsoft NET and J2EE: Microsoft(r) .Net and J2ee*. s.l.: Microsoft Press, 978-8120326682.

Morris, Jr., John B. 2011. "Injecting the Public Interest into Internet Standards." In *Opening Standards: The Global Politics of Interoperability* (The Information Society Series), ed. Laura DeNardis. Cambridge: The MIT Press.

Panetto, Hervé, Monica Scannapieco, and Martin Zelm. 2004. "INTEROP NoE: Interoperability Research for Networked Enterprises Applications and Software." In *On the Move to Meaningful Internet Systems 2004: OTM 2004 Workshops*, eds. Robert Meersman, Zahir Tari, and Angelo Corsaro. Berlin: Springer/Heidelberg, 866-882.

Roberts, E. and P.S. Gallagher. 2010. "Challenges to SCORM." In *Learning on Demand: ADL and the Future of e-Learning*, eds. Jesukiewicz P., Kahn B., and Wisher R. Alexandria, VA: Advanced Distributed Learning.

Roman, P. and R. Bassarab. 2008. The Simulation Interoperability Framework—Mapping Interoperability Requirements to Coalition Outcomes, NATO.

Ryan, M., D. Hill, and D. McGrath. 2005. "Simulation Interoperability with a Commercial Game Engine." European Simulation Interoperability Workshop 2005, 27–30 June 2005.

Spires, H. 2008. "21st Century Skills and Serious Games: Preparing the N Generation." In *Serious Educational Games: From Theory to Practice,* ed. L.A. Annetta. s.l.: Sense Publishers, 13-24. Please provide place of publishing in ref. Spires (2008).

Stănescu, I.A., I. Roceanu, A. ªtefan, and I. Martinez-Ortiz. 2011. "Principles of Serious Games Interoperability." In *ISI Proceedings, the 6th International Conference on Virtual Learning*, Cluj Napoca, Romania, October 28–29.

Stănescu I.A., A. ªtefan, and I. Roceanu. 2010. "Serious Games Interoperability." In *ISI Proceedings, the 7th eLSE Conference*, Bucharest, Romania, April 15–16.

Aspects Of Serious Games Curriculum Integration — A Two-Fold Approach

Maria-Magdalena Popescu; Ion Roceanu
Carol I National Defense University, Romania

Jeffrey Earp; Michela Ott
Istituto per le Tecnologie Didattiche, Consiglio Nazionale delle Ricerche, Italy

Pablo Moreno Ger
Complutense University in Madrid, Spain

Abstract

Over the years, there have been numerous definitions of curriculum integration, where the curriculum is interwoven, connected, thematic, interdisciplinary, multidisciplinary, correlated, linked, and holistic (Fogarty and Pete 2007). *Curriculum integration is based on both philosophy and practicality, drawing together knowledge, skills, attitudes, and values from within or across subject areas to develop a more powerful understanding of key information. Curriculum integration is best done when components of the curriculum are connected and related in meaningful ways by both students and teachers. With the large uptake of SGs in education nowadays, one must consider SGs curriculum integration an issue at large since effectiveness of SGs use in training and education is getting more and more proponents. This paper looks at SGs curriculum integration issues from two perspectives—of the teacher connecting the content of the game and the learning outcomes into the whole educational context on the one hand, and of the researcher who sees the connection between the pedagogical state-of-the art in SG and what realia can offer, on the other. By drawing on the experience of three teams of researchers and educators from Romania, Italy, and Spain, based on common activities conducted by same partners and others in the Games and Learning Alliance (GaLA), an EC-funded Network of Excellence on SGs, joint perspectives over curriculum integration will be presented, with a view to sharing the experience in order to give guidelines for a future extension of SGs into education and training, into well-built curricula. The situations presented of SGs curriculum integration in three different educational contexts are to showcase the framework for building an SGs curriculum design, the way SGs are effective for instruction, to present forms of integrating an SG into the curriculum—how, where, how long, and showcase trans- and inter-disciplinarity within SG curriculum integration. A set of guidelines will be just a quick overview on what both practitioners, researchers, and policymakers should consider for the near future in terms of SG currriculum integration, to enhance a large-scale uptake of SGs into all levels of education and training, to better respond to the twenty-first century student and current social needs. All of the statements and observations will be outspoken based on genuine results of the experiments and long-term practice of the authors in the realm of SGs integration into the training programs.*

KEY WORDS: *serious games, curriculum integration, pedagogy, psychology*

I. Curriculum Integration As A New Pedagogical Approach

An educational system is never an island; on the contrary, it is a continuum throughout society, culture, politics, economics, and everything a country builds its foundation on. An educational system takes after the society it is built within and for which it matters at present. Attracting individuals into coming to school especially if adult education is involved is a permanent challenge nowadays, where consumerism and popular culture are at their best. More than children, Higher Education and Further Education students have clearly cut and timely framed objectives. They are attracted to learning if they receive what they look for. Student books and teaching aids are less relevant unless highlighted and integrated into a carefully designed curriculum, with a valuable content, highly applicable in the contingent reality.

Not very long ago we only spoke of Net generation or the New Millennials, these young people whose fingers restlessly lay on the computer devices as naturally as possible, have grown up with computer games (Oblinger and Oblinger 2005) and have already turned to Higher Education while the older generations have done their best to adapt to high-standard requirements. Technological advances and serious games growing uptake as complementary teaching tools, their proven educational effectiveness in the training process have given rise to a re-think of the curriculum and of the learning and teaching paradigms.

1.1. SG integrated inside curricula

Using serious games in education in a perpetual need to meet the ever growing requirements of a multi-skilled individual in a multi-cultural, multi-faceted society and labour market, asks for new pedagogical approaches in game-enhanced learning: the one we chose to tackle here, curriculum integration, is a teaching approach that enables students and teachers to identify and research problems and issues regardless of subject-area boundaries. "The very notion of 'integration' incorporates the idea of unity between forms of knowledge and the respective disciplines" (Pring 1973)

Curriculum Integration basically covers real-life themes enabling students to be inquisitive and pragmatic for real-life issues, to collaborate with their peers and teachers as well, it unifies learning related to subject areas and has students use an inordinate number of skills to inquire on present-day, living concerns, on combined disciplines of study. Moreover, students benefit from wide knowledge-access by means of a relevant learning process, irrespective of their backgrounds and abilities.

Speaking about an interdisciplinary curriculum, Loepp (1999) considered that this can be closely related to an integrated curriculum while educational researchers have found that an integrated curriculum can result in greater intellectual curiosity, improved attitude toward schooling, enhanced problem-solving skills, and higher achievement in college.

Serious Games or game-based learning in general is to curriculum integration what hand is to glove.

In this respect, by playing for example Quest Atlantis, participants in this game will develop problem-solving skills, decision making, affective skills, based on previous knowledge on biology, physics, art, social sciences (build shelters and foster creativity), environmental issues (considering scientists who analyze data about water quality to diagnose why fish are dying), and demographics (students must choose between renovating a homeless shelter and building a park). Similarly, in Civilization III students have to integrate knowledge on history, economics, foreign policy, and geography, as "a form of transgressive play" (Squire and Jenkins 2004). If designed correctly from the outset, a game can successfully integrate more subject-matters within one and the same context, in a trans-disciplinary way, thus touching upon both cognitive, affective, inter and intra-personal skills. The problem remains for the commercial-off-the-shelf games that are already in use, for cost-effective reasons. Is this a matter of one size fits all? Can they be used in a game-based—learning—curriculum integration approach or are there alterations to be made in terms of pedagogical approach and tutor's role? Is this situation much different from the specially designed or modding games and how far do the implications go in relation to the educational environment and training effectiveness, the areas covered in subject-matters and variety of skills to be developed or reinforced?

II. Practitioners' Viewpoint

There have been inordinate studies and practices on integrating games into curriculum, focusing on the appropriateness of such an initiative, both considering the origin of the game and the way they respond to the envisaged learning objectives. Debates whether it is more effective to use COTS or build games from scratch either by students themselves or by teams of educators coordinated by game designers have filled pages of conference proceedings. While using commercial off-the-shelf games means taking up games the way they are, not necessarily developed as learning games, and using them in the classroom, one must consider that not all games are designed to teach, hence the subject matter taught may hardly find common points to the game and the content could be far from complete in relation to the things taught (Van Eck 2006). Conversely, building games from scratch to answer certain curricula might again be inefficient as by the time a game is developed the curriculum might change and then once ready, we can just discover the game needs improvement again, to correspond to the newly designed requirements. Hence, a careful analysis to match the contents of the game to what has to be studied *can only be obtained in a careful analysis of the game prior to its implementation.*

2.1. Time management, pedagogies, uses, and drawbacks

Time savings could be properly obtained if full guides of COTS games provided enough discrete information on the story, contents, and possible learning objectives to be met in case of game use. However this is immense work and the dusk of it is just here— "serious games classification" repository and IMAGINE as well as ENGAGE provide basic descriptions of games, lacking though important descriptors like duration of game per sequence or per full learning process; if the game can be used as guided practice, as reinforcement or development of certain skills, if the game offers procedural or declarative knowledge, if it assesses or even if it can be used as transdisciplinary project-based evaluation, or mere incentive for theoretical approach on a single subject-matter. These, along with targeted audience and any pre-requisites for learning would help the decision maker—in the person of an educator, a policymaker, or corporate training stakeholder—to select the most appropriate games and implement them properly inside the curriculum.

Once these instruments are at hand, then the games integration process will follow the gauntlet track of any curriculum development model, course design—the course/courses that will actually embed the game: aim, content, teaching, and learning methods; there are yet cases where the game is embedded inside the syllabus only, depending on the game content and the possibilities the latter offers for exploitation.

Moreover, theories of adult learning, student centred learning, active learning, and self-directed learning may all influence the overall programme philosophy, similar to other elements—student needs such as the need for flexible learning programmes (McKimm 2003). With respect to this situation the idea of breaking down the game—if need be—into sequences corresponding to the learning modules might be needed. Here, the idea of lowering the entertainment aspect of a game may be brought into discussion.

Once the curriculum is built on the premises of an Outcomes Based Education (OBE) which states that "educators should think about the desirable outcomes of their programmes and state them in clear and precise terms" (Prideaux 2000) they should then work backwards, to determine the appropriate learning experiences which will lead to the stated outcomes. By using an outcome approach, educators are forced to give primacy to what learners will do and to organize their curricula accordingly—is also what Prideaux considers.

Similarly, a balance between the needs of the curriculum and the structure of the game must be achieved to avoid either compromising the learning outcomes or forcing a game to work in a way for which it is not suited" (Van Eck 2006).

The way the game is then incorporated into the lesson itself once projected into the syllabus is just the educator's say. The way he makes students feel responsible for going through the game as a continuum to the real-life situation or sometimes as a pre-requisite for real-life—like activities within the learning process is only given by the methodologies he uses. Differences must be made here though among the K-12, HE, and adult education pedagogies to maximize the use of the game to answer the vast array of students' needs, interests and style.

III. Games Seen As Integrated Curriculum
The introduction of games in school curricula represents a key novelty for most EU countries (Mitchell and Savill-Smith 2004). Things differ in approaches where COTS or games designed from scratch are used. Thus, in case of games developed from scratch, accurate design and careful planning are required, together with the adoption of new educational approaches (Kirriemuir and McFarlane 2004).

3.1. Researcher's point of view
To later introduce them within the educational process, pilot experiments in the field are often carried out by composite teams where researchers (e.g. researchers in Educational Technology and Educational Psychology) join and assist school teachers with both the aim of sustaining the experiments from a theoretical point of view and of acquiring data from the experience to better tune future interventions, related models, and methods. The type and level of the actual collaboration between the research world and the school world varies a lot. In most situations, researchers perform both the role of devising and designing the educational actions to be carried out; they then inform and appropriately train teachers; teachers, on the other hand, are often the only ones commissioned to conduct the school experiment and to gather sensible data (via questionnaires that researchers have provided); at a later (often at the last) stage of the experiment researchers study and analyze the available results, perform the final evaluation of the experiment, draw the related theoretical and practical conclusions, and publish the sensible outcomes.

Conversely, a different approach to collaboration between researchers and teachers is also possible and, in our opinion, can offer significant added value to reaching the entailed educational objectives (for schools) and the effectiveness and repeatability of the conducted experiments (for research institutions).

This triggers a more direct and mutual collaboration of researchers and teachers in all of the four basic stages of a field experiment (Figure 1):

Figure 1. Phases of a game-based learning field experiment

This more "close and inclusive" collaborative approach between researchers and teachers was adopted by ITD-CNR in Italy while carrying out a pilot game-based experiment (Bottino and Ott 2006; Bottino et al. 2007) in primary schools aimed at supporting and triggering young students' cognitive abilities by means of games deeply requiring the enactment of reasoning and logical skills.

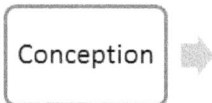

The idea itself of conducting this type of experiment actually originated from the dialogue between teachers and researchers in Educational Technology: the former asked for some kind of ICT-enhanced tool which was able to sustain the children's reasoning abilities and the latter, based on previous research projects on the use of COTS games, imagined that such games could profitably be used for the intended scope. Common reflections of the two types of actors led to deciding which types of games were more appropriate: teachers, for instance, pointed out that games presenting no interference with other curricular abilities (e.g. arithmetic) would represent a better solution in order to help children concentrate on the reasoning tasks; researchers, on the other side, individuated the most appropriate tools based on their specific knowledge of software dynamics, software interface, and game mechanics. Hence, following some teachers' observations, experts in special needs education were included in the research team to better understand and address encountered needs and specific problems.

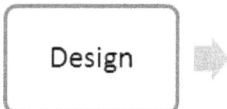

The contribution of teachers was also very important during the phase of designing the overall educational intervention, normally demanded only of researchers. As a matter of fact, teachers have a more precise idea and in-depth knowledge of the peculiarities present in each class main characteristics and for each student; in particular, they also know the specific setting where the experiment will take place and the time that can be allocated to it. In the case at hand, teachers gave a sensible contribution to the planning of activities both as to the general contextual aspects and also to those related to contents and possible personalization of the educational paths.

Enactment

The enactment phase, the one where children played the games, fully demanded that both teachers and researchers play a significant role: to follow the students during the gaming sessions (Figure 2), together with the teachers and the special needs educators: this allowed a multifaceted monitoring of the situation, revising the fine tuning, amending, and improving the monitoring and evaluation sheets that had been "ad hoc" produced to allow data retrieval and analysis.

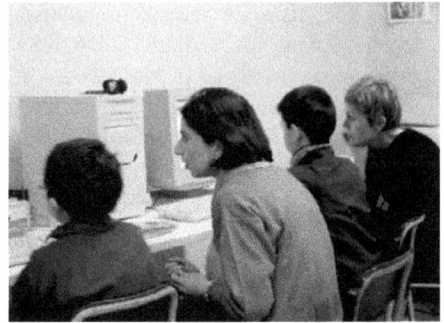

Figure 2. Aspects of the joint work teachers–researchers during experiments: sharing decisions and monitoring students

Evaluation

The genuinely common work carried out in the previous phases of the project by the full crew had a particularly important impact on the evaluation phase. Although in this phase researchers (educational technologists and psychologists) were in charge of elaborating data and carrying out statistical data analysis, the overall evaluation of available data highly benefitted from the contribution of all of the team members. The gained insight in the students' learning process would not have been so in-depth and so effective without the single contribution of each team member. Each of them could give his own contribution from his specific stand point but having personally participated in all of the intermediate steps of the learning process, he/she was able to frame it in the general context of the overall experiment, thus coming to consistent and homogeneous conclusion with the others.

IV. Points To Consider Within Curriculum Integration
As was showcased in the previous sections, serious games can complement the standard rigid curriculum of schools, providing transversal learning activities as well. Unlike the already mentioned situation where games that had no interference with other curricular activities were used, the basic idea would be to complement the classes with

special sessions in which the students play serious games and reflect upon their play, linking their different in-game activities with different areas of the curriculum.

However, in spite of the important effect achieved by initiatives such as the one described in the previous section, the adoption of serious games in general education remains slow and elusive. To begin with, there are social, cultural, and technical barriers that hinder real application of videogames in schools. For each successful case study, there are dozens of failed attempts to convince schools to explore serious games-based learning approaches. In this section we summarize some of the barriers and challenges of bringing games into the educational process in general, and how a curriculum integration approach could overcome them.

4.1. Bringing a game into the classroom: barriers and challenges

One of the main battlefronts when bringing a game into a school is the much degraded social perception of videogames. Media coverage tends to focus exclusively on controversial games, and if we were to study the medium only through its coverage on TV, we would deduce that all games are extremely violent, that they all purport explicit pornography and that only lonely male kids play them.[1] In this context, it is normal that parents and teachers display great concern when the idea of introducing games into the classroom is presented.

In addition, parents legitimately argue that using games in class as part of the curriculum may undermine their ongoing efforts to control the time their children spend playing games at home.

Moreover, since games are an advanced form of technology, they will undergo the gauntlet any innovation goes through: Teachers tend to resist innovation, especially when such innovation may be a drawback against their more tech-savvy students. Even supposing teachers would accept that and turn themselves into guides (rather than oracles), the availability of proper technologies in many schools is scarce, giving rise to hindrance of the process as well.

In addition, the syllabus is tight as it is, and there is little time for any kind of extraordinary activity. In this respect, educational authorities in many countries have made an effort to provide public schools with adequate IT infrastructure to introduce a more relaxed and effective curriculum, yet school staff still lack proper training to use them.

Finally, even students have been found to reject game-based learning approaches, rapidly identifying them as requiring more time and effort and preferring minimal effort approaches (Squire and Barab 2004).

4.2. Bringing the game out of the classroom

All of the barriers and resistances presented by schools end up discouraging further growth and research in this area, and it seems obvious that schools are not prepared to embrace game-based learning as part of their curricular activities. Christensen et al. (2008) discussed the notion that schools entrenched in fixed methodologies cannot take big steps forward. Drawing from experiences in other industries, they suggest that

---

[1] This contrasts with the data gathered by the Entertainment Software Association, which claims that the average age of videogame players is 34 and that 43% of them are female.

advances in education should be disruptive, targeting niche markets first and then growing from there.

Their suggestion focuses on the use of student-centered online education systems that cater to the needs of each individual student, as a means to customize the learning experiences beyond the ability of formal schooling. From their perspective, as these customized learning experiences grow, they will eventually gather enough momentum to compete with the traditional school model, or even displace it entirely.

This very same idea has been always at the core of our research with educational games. We believe that game-based learning can thrive in the more open-ended and innovative e-learning arena, and eventually use that as a vehicle to enter the schools (Moreno-Ger, Burgos and Torrente 2009), and this is already happening. Higher education institutions are massively embracing blended-learning models, in which traditional face-to- face lectures are complemented (rather than substituted) with e-learning technologies (the so-called, *learning management systems (LMS)*). Using these systems as the infrastructure to deploy educational games allows their employment today as a complement in higher education settings and tomorrow, hopefully, in all kinds of educational settings.

4.3. Blended curriculum integration

In this view, it would be possible to explore new educational models that attempt to complement the traditional curriculum with integrating activities, rather than overhauling the existing curriculum to make it integrated. The key idea would be that students can play at home, on their own computers or gaming devices, in direct connection with the school's online LMS.

This connection would allow the integration of game outcomes with the other curricular activities' outcomes and assignments, while the learning management system (LMS) would connect the game sessions at home to the reflection sessions in school.

Hence, the games can act as transversal curriculum integration activities, while the teachers act as facilitators of the process by connecting those activities with the regular curriculum: students would play at home and then participate in debriefing sessions at school, facilitated by the teachers. These debriefing sessions not only relate the game to the content, but also enhance reflection about play, as important as play itself (Peters and Vissers 2004).

Yet, in order to facilitate this debriefing process, the facilitators need insight into the students played the game, along with the certainty of having done that. This requires having games that can track the gameplay session and create feedback reports used to guide those debriefing sessions.

From this perspective we created eAdventure[2] as a tool to facilitate the creation of games as a complement to education in blended learning environments. eAdventure is an authoring platform for the creation of educational games that tries to overcome some of the challenges of educational gaming highlighted above. The games created with this editor can be run either as a stand-alone tool (allowing instructors or learners to execute educational games on their computer) or embedded in a web-based e-learning system.

The games include features to track the movements of the player, and it is possible to create assessment reports that summarize the most meaningful events from a learning perspective (as indicated by the instructor). These reports can take the form of a human-readable log or send data to be stored in the system. This is achieved using

---

[2] http://e-adventure.e-ucm.es

the communication APIs described in the SCORM framework or through one of the ad-hoc communication mechanisms supported by the platform (del Blanco et al. In press).

V. Conclusions And Ways Forward

The joint work of teachers and researchers requires, as previously highlighted, the capacity of selecting appropriate tools and resources in accordance with specific learning objectives and of devising appropriate and suitable educational methods. The educational effectiveness of games (as well as that of any other technological means) mainly depends on the choices made by those in charge of designing and setting up the activity: in order to take a significant step forward, the use of e-tools needs to be carefully planned and structured, and conceptually well integrated in mainstream activities, bearing in mind that e-tools (including digital games) do not make the difference per se, simply by being used; it is the concepts and ideas underpinning the learning activities that produce effective and significant changes on educational processes and the related pedagogical planning.

Thus, regarding the use of integrating game activities outside the classroom as a complement to traditional education, the key challenge is their meaningful integration so that it is possible to connect the game activities with different aspects of the traditional curriculum.

In this sense, we have suggested using an online LMS to deploy the games so that students can play from home, at their own pace, on their own computers. In order to avoid this "playing at home" from becoming a barrier for posterior reflection and debriefing sessions, it is necessary to produce games that provide insight into how the game was played by each student. Traditional LMSs track when each student accesses each piece of content, but simply knowing that the student did open the game at home is not enough for preparing a meaningful debriefing session. Games should include mechanisms to track the activity of each student *inside the game*, providing insight into where the students stumbled, found problems, or tried different things. The eAdventure platform, as described in Section 4.3, represents a first step toward meaningful integration, proposing a model in which the games report back to a central server using standard-compliant communication methods.

It is yet worth mentioning that while games will certainly not replace the teacher, as some fear, they can open the way to more creative approaches that could have a significant impact on teaching practices (Popescu et al. 2011), by simply engaging transversal learning where more skills are challenged into project based-type of activities, fostering not only cognitive, but also motor and affective skills similarly, provided they are well chosen in accordance with the subject-matters they can refer to as a continuum, within the syllabus or—on a larger basis—within the curriculum.

Moreover, from the stakeholder's viewpoint, numerous education institutions, particularly universities and colleges, have identified advanced distributed learning as the first priority of their development strategy. The importance and expansion of this kind of education has grown in the last year at a pace that shows the feasibility of the modern education system created in recent years (Calopareanu 2011) setting, thus the proper environment for alternate means of instruction and teaching devices among which Serious Games are a distinctive category based on the inordinate challenges they bring and the novelty toward making learning and real-life application a continuum from which both students and instructors benefit, along with labour market stakeholders to a final end.

About the Authors
Maria-Magdalena Popescu is an associate Professor in the Foreign Languages Department at the "Carol I" National Defense University, in Bucharest, Romania
Ion Roceanu serves as Director of the Advanced Distributed Learning Department at the "Carol I" National Defense University in Bucharest, Romania.
Jeffrey Earp is a research assistant at the Institute of Educational Technology, Italian Research Council (ITD-CNR).
Michela Ott is a senior researcher at the Institute of Educational Technology, Italian Research Council (ITD-CNR).
Pablo Moreno Ger is an associate professor of computer science at the Complutense University of Madrid.

References
Bottino, R.M., a n d M . Ott. 2006. "Mind Games, Reasoning Skills, and The Primary School Curriculum: Hints From A Field Experiment." *Learning Media & Technology*, 31 (4): 359-375.
Bottino, R.M., L . Ferlino, M . Ott, a n d M . Tavella. 2007. "Developing Strategic and Reasoning Abilities with Computer Games at Primary School Level." *Computers & Education* 49 (4): 1272-1286.
Calopareanu, Gheorghe. 2011. *Current Trends in ADL.* http://adlunap.ro/eLSE_publications/ papers/2011/1699_1.pdf at 27.02.2012.
Christensen, C., C.W. Johnson, and M.B. Horn. 2008. *Disrupting Class: How Disruptive Innovation Will Change the Way the World Learns.* McGraw-Hill, 288.
del Blanco, Á., J. Torrente, E.J. Marchiori, I. Martínez-Ortiz, P. Moreno-Ger, and B. Fernández-Manjón. In press. "A Framework For Simplifying Educator Tasks Related to the Integration of Games in the Learning Flow." *Education Technology & Society*.
Fogarty, R.J. and J.B. Pete. 2007. *How to Differentiate Learning, Curriculum, Instruction, Assessment,* Corwin Press
Kirriemuir, J. and A . McFarlane. 2004. *Literature review in games and learning,* [online], Futurelab Series, Report 8, UK. http://www.nestafuturelab.org/research/reviews/08_01.htm.
Loepp, F.L. 1999. "Models of Curriculum Integration," *The Journal of Technology Studies*
McKimm, J. 2003. *Curriculum Design and Development.* http://www.faculty. londondeanery.ac.uk/e-learning/setting-learning-objectives/Curriculum_design_and_development.pdf.
Mitchell, A. a n d C . Savill-Smith. 2004. *The Use of Computer and Video Games for Learning,* UK: Learning and Skills Development Agency. www.LSDA.org.uk.

Moreno-Ger, P., D. Burgos, and J. Torrente. 2009. "Digital Games in eLearning Environments: Current Uses and Emerging Trends." *Simulation & Gaming* 40 (5): 669-687.

Oblinger and Oblinger. 2005. *Educating the Net Generation*, EDUCAUSE.

Peters, V.a.M. and G.a.N. Vissers. 2004. "A Simple Classification Model for Debriefing Simulation Games." *Simulation & Gaming* 35 (1): 70-84. doi:10.1177/1046878103253719.

Popescu M., S. Arnab, R. Berta, J. Earp, S. De Freitas, M. Romero, I. Stanescu, and M. Usart. 2011. *"Serious Games in Formal Education: Discussing Some Critical Aspects."* In: *ECGBL 2011—The 5th European Conference on Games Based Learning. Proceedings*, eds. Dimitris Gouscos, Michalis Meimaris. Academin Publishing Limited, 486-493.

Prideaux, D. 2000. "The Emperor's New Clothes: From Objectives to Outcomes." *Medical Education* 34: 168-169.

Pring, R. 1973. "Curriculum Integration." In *The Philosophy of Education*, ed. R.S. Peters.. London: Oxford University Press, 123-149.

Squire, K. and S. Barab 2004. "Replaying History: Engaging Urban Underserved Students in Learning World History through Computer Simulation Games." In *6th International Conference on Learning Sciences*. Santa Monica, United States: International Society of the Learning Sciences, 505-512.

Squire, K. and H. Jenkins 2004. "Harnessing the Power of Games in Education." *Insight* 3 (5): 7-33.

Van Eck, Richard. March/April 2006. "Digital Game-Based Learning: It's Not Just the Digital Natives Who Are Restless." *EDUCAUSE Review* 41 (2).

Serious Game Application In Anti-Aircraft Missile Training

Silviu Apostol; Fabian Breharu
Insoft Development and Consulting, Romania

Florin Mingireanu
Romanian Space Agency

Abstract

A serious game application related to anti-aircraft missile training is presented together with strategies specific to the field. Technologies used to develop the application together with ballistic models are presented in comparison with real-life training applications. The application allows the user to train using several pre-set cases for specific aerial targets. Through specific e-learning strategies the user capability and experience is enriched together with specific automatism directly related to the field of anti-aircraft missile training. Key advantages of the applications are presented among which we mention low-cost, repeatability capability, detailed analysis of missile firing tests, as well as detailed equipment resemblance.
The interface is realized with great detail and special care is given to the real-life feeling of various buttons and switches. The application tries to give to the trainee a real-life feeling with smooth performance capability.
Interaction with various equipment boards is enabled through easy mouse-click interaction.
Within the application the user has the freedom to choose between training and the real fire case. In the training the user gets used to the equipment and the computer indicates through specific labels the specific steps that the user should perform at any stage. Limited action is enabled and the user cannot be make a mistake as he/she must be focused on learning the correct button position for specific targets.
Within the real fire case the user is offered by the computer a random target and he/she must take the correct decisions and must set the correct buttons for the specific target. Here, he/she must use the knowledge acquired within the training mode of the application.
The entire application is set within a game scenario and should offer the users the capability to train
repeatedly at lower cost than real-life fire tests. Combined with the real-life fire tests this application enriches the user experience and expertise within the area of anti-aircraft missile operation. The immersion offered through both graphical and algorithm design guarantees excellent transfer from the game frame to the real-life frame for the typical user of anti-aircraft missile technology.

KEY WORDS: *defense, missile, air, radar, rocket, elearning*

I. Chapter I

Air defense systems have been used to maintain air authority over a given airspace. With the advancements of technology, both radars and guns/missiles have become increasingly more efficient. On the one hand, radars, started to have a longer range and a lower surface detection threshold while on the other, guns'/missiles' ranges increased and were more accurate.

Most of the advancements were performed during the Cold War when both the USA and the USSR developed many air defense systems. Developments of transistors and, later, of integrated circuits and microprocessors offered the possibility in increased performance of radars and computers, opening new possibilities for worldwide air defense systems.

At the same time, a number of great advancements were made in the field of missiles. Among these advancements web were the development of solid rocket motors, liquid rocket motors, new alloys, increasing manufacturing precision through the use of automatic machines, command and control systems as well as warhead improvement.

Due to the strategic conditions, the USSR designed and developed a large number of air defense systems among which we mention: SA-4 Ganef (Fig. 1), SA-19/SA-N-11 Grison (Figure 2), and S-75 Volhov (Figure 3).

Figure 1. SA-4 Ganef

Figure 2. SA-19/SA-N-11 Grison

Figure 3. Volhov missile system

The S-75 Volhov system was the most widely deployed system covering countries from four continents. This system was a high-altitude, command-guided system intended to combat aerial targets flying at high altitudes and at high speeds. It scored the first destruction of an enemy aircraft by a Surface-to-Air-Missile (SAM), shooting down a Taiwanese Martin RB-57D Canberra over China, on October 7, 1959 by hitting it with three V-750 (1D) missiles at an altitude of 20 km (65,600 ft.).

Many of the countries that acquired the S-75 air defense system had invested in various training programs that included firing exercises both in the country as well as in dedicated fields within the USSR. This training involved great expense aside from the cost of the system itself however the training was necessary because otherwise the efficiency of the system would have been very low. This lack of training was observed especially during the Vietnam War when Vietnamese operators were not able to shoot down a plane even when six missiles were fired upon that specific plane. In contrast to this, a well-trained Soviet operator could bring down the plane with no more than three missiles fired; in some cases, even one missile was sufficient given that the operator was well trained. The decision of when to fire the missile, how to fire it, what guidance program to use, what fuse mode to use, and other details were usually learned in a crash program that was more or less efficient depending on the capacity of the operators to learn under a fast pace.

Nowadays, computer technology and e-learning has opened up a new capability to create "serious games" intended to train a specific person for specific skills that he/she has to perform within the work environment. Our paper presents a serious game application developed to train operators of air defense systems. We chose the S-75 as an example due to its wide availability in the world. However, the application can be adapted for any other type of air defense system.

II. Chapter II

The Volhov serious game application was developed using Flash technology that can run under a browser on a computer with minimum 2 GHz processor, 1 Gb RAM and 1 Gb free hard disk space.

The Volhov project was made with Adobe Flash CS5 for its ease of implementation of graphical aspects. The simulation is separated into different types of modules that work independently and one module that correlates the actions of the user with the simulation. For example, the user interface (starting menu and the panels) is managed by one module which sends notifications to the control module when something needs to be updated. Due to simulation considerations, the radars, enemy planes, and missiles are also different modules, thus providing a behavior close to the real missile system since the missiles' (and enemy targets) only goal is to reach their target and the radars' only goal is detecting an enemy threat (if the user manages to switch on the radars) and forwarding the information to the main panel monitoring equipment for altitude, distance from base, and velocity. Of course, if the generator is switched off, the panels become inoperable, but do keep track of button presses so when the generator is switched on, the changes are displayed accordingly. All of this is then correlated with the graphical user interface that uses vector drawings and a minor 3D effect when switching from one panel to another. So for instance, if you launched a missile toward an enemy plane and you decided to view the missile at any stage in its flight, the "outside view" module would be updated regarding the status of the simulation within a minor time interval (for example if you launch the missile, you will be able to see the launch even if you were 5–6 seconds late in clicking the "outside view" button and you will also be able to see missile hits/misses).

The application was structured in two modes:
- Training
- Firing exercise

In the training mode the operator is shown tooltips with relevant information on what to do next and why he/she should do it. Basically, in this mode the operator familiarizes himself/herself with the buttons and panels of the real system. All of the panels and buttons were built using high-resolution photos of the real equipment embedding them in an easy-to-use graphical interface. In the training mode the operator is presented with several types of targets (three types) and for each type he/she is taught how to combat that specific type and what configuration to setup on the panels in order to maximize the chances to combat that target.

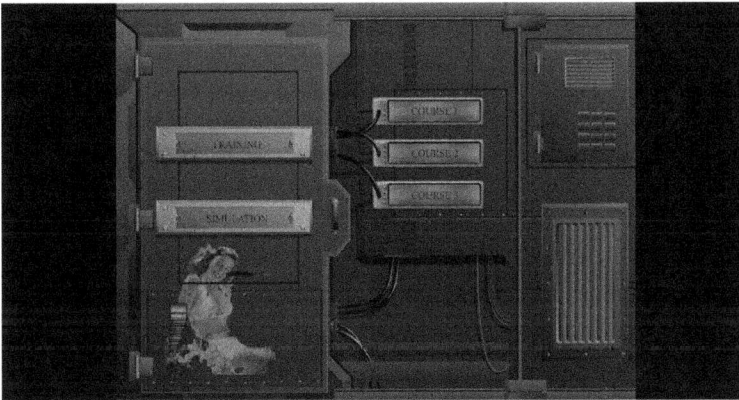

Figure 4. Simulator modes

In the firing exercise the operator is randomly presented a target and he/she should perform all of the needed tasks without receiving any assistance from the computer. In this mode, the operator proves what he/she learned in the training mode. Depending on the type of target (high/low altitude, high/low speed) the operator has a certain amount of time within which he/she has to complete all of the configurations on the panels and fire 1/2/3 missiles against the target.

Figure 5. Command and control radar

The time limitation is yet another realistic feature of the application since it adds the time factor, which is very important in a real situation. In other words, even if the operator performs all of the configurations needed for a specific target, he/she must also perform the configuration within a limited time. If he/she goes over the time then the game is lost because the target reached the air defense location before it was destroyed. In reality, this is equal to the destruction of the air defense base or of the mission objective that the target might have had.

Two of the most important parameters for serious game applications are the following (Werkhoven and Van Erp; Millán et al. 2004; Friedman et al. 2007)

(1) Validated content
The content is very important and the models on which the content is based should be as realistic as possible. In order for people to experience causal relations in concept testing it is of crucial importance to develop, validate, and combine models that define the behavior of the action-response patterns in the simulated world. Some examples of various categories are models of the physical, cognitive, and group behavior of virtual characters, public governance models, dispersion models of chemical and biological warfare agents, models of the explosion sensitivity of built constructions, and models of interdependencies within the critical vital infrastructure, etc. Also, it is intended that the content be validated in greater detail through collaboration with the Technical Military Academy from Bucharest, Romania.

(2) Intuitive while realistic interfaces (Taylor et al. 2006)[5]
As with any other computer application, an intuitive interface is needed in order to gain maximum knowledge from the content presented. More than being intuitive, the interface should be as realistic as possible and provide the user, especially in the training mode, as many indications as possible. However, the interfaces should be kept as realistic as possible; otherwise the danger is that the trainee might not recognize the specific equipment in reality. In other words, the interface should give the trainee the possibility to immediately recognize the components of the real device.

Future plans for developing this application include an extension of the interface toward 3D as well as a closer to reality resemblance of all graphical components. An extension of the algorithm behind the guidance of missiles is considered in order to offer more realistic ballistic characteristics.

All of these improvements would create an increased immersion as well as a better training performance for the operator.

III. Recommendation
Through the use of computers, realistic training applications for air defense systems can be developed. The costs of training are reduced because the operators need less actual firing hours at the firing range thus providing increased efficiency of the training process.

If, in reality, an error means a lost missile, on the computer this can be repaired by simply resetting the application. The operator can repeat the firing many times over and the training of automatic reflexes are attained. For an air defense operator, having automatic reflexes is a must. During a real firing exercise or during real combat there is no time to "think it over"; you must act and the action should be automatic. The difference between winning or losing a battle is obvious if the operator is not correctly trained and correct automatic reflexes are not achieved.

Acknowledgments
The authors would like to thank Ionel Fertu for providing some details of the system and INSOFT Development and Consulting for providing the technical capabilities to produce the demo application as well as graphical components.

The authors would also like to thank the Technical Military Academy from Bucharest, Romania for the overview and feedback they provided regarding the anti-aircraft missile simulator.

About the Authors
Silviu Apostol is the instructional design manager at InSoft Development and Consulting in Romania, as well as a biologist and instructional designer at the University of Bucharest.
Fabian Breharu works at Insoft Development and Consulting, Romania.
Florin Mingireanu is an associate researcher at NASA and the Romanian Space Agency.

References
Friedman D., R. Leeb, C. Guger, A. Steed, G. Pfurtscheller, and M. Slater. 2007. "Navigating Virtual Reality by thought: What is it like?" *Presence* 16 (1): 100-110.

Millán, J. del R., F. Renkens, J. Mourino, and W. Gerstner. 2004. "Noninvasive Brain-Actuated Control of a Mobile Robot by Human EEG." *IEEE Trans Biomed Eng* 51 (6), 1026-1033.

Taylor, D., W. Huiskamp, K. Kvernsveen, and C. Wood. 2006. "Preliminary Analysis of Tactical Data Link." "Representation in Extended Air Defense Simulation Federations." *Proceedings of Simulation Interoperability Workshop SIW*, Huntsville, USA.

Werkhoven, Peter J., and Jan B.F. Van Erp. "Serious gaming requires serious interfaces". http://hmi.ewi.utwente.nl/brainplay07_files/werkhoven.pdf

Students' Time Perspective And Its Effect On Game-Based Learning

Mireia Usart & Margarida Romero
ESADE School of Law and Business, Spain

Abstract

Previous research in face-to-face learning modality demonstrated that students' Time Perspective (TP) is related to motivation and learning performance. Concretely, results show students with a future-oriented TP having higher motivation for learning, higher self-regulation, and academic performance. By contrast, students' having a present-oriented TP tend to engage in games and prefer instant reward activities. Despite the wide corpus of research on TP and learning, albeit Serious Games (SG) are widely used for professional development and lifelong learning, no studies have focused, as per our knowledge, on TP in Game Based Learning (GBL). The present study aims to explore this new field of research. We conducted a case study using the Serious Game MetaVals. Results of the experience show no significant differences in game performance among individuals with different TP. Furthermore, students with a future-oriented TP foresee the future usefulness of the game compared to those focused on the present. These results might be useful for instructional designers and teachers, in terms of knowledge acquisition, outlining the benefits of using GBL activities that could help different TP profiles to equally engage and better perform in the learning processes.

KEY WORDS: *Game Based Learning, Serious Games, Time Factor, Time Management, Time Perspective, Learning Performance*

I. Introduction

Continuing professional development and lifelong learning are vital to both individual and organizational success (Wall and Ahmed 2008). Present trends in management education are committed to active learning models, including Serious Games (SG), in their curriculums. Especially, Game Based Learning (GBL) has long been used for management training courses, to safely practice key skills and competencies in students' improvement (Mawdesley et al. 2011). Furthermore, the time factor plays an important role in these new learning scenarios (Gros, Barberà, and Kirshner 2010): students have to be aware of the existing time constraints in their life, and therefore manage time to take advantage of their learning process. This study aims to analyze a specific aspect of the time factor, namely Time Perspective (TP). We analyzed students' TP in relation to learning performance, intention of use, and usefulness of MetaVals, in SG on finance basics.

This study was developed within the context of a PhD, focused on adult students' Time Perspective (TP) and its possible effects on GBL activities, conducted in ESADE Business and Law School. The study is set within the Network of Excellence FP7 Games and Learning Alliance (GaLA), in the Special Interest Groups of Pedagogy and Psychology.

1.1. Game based learning

The use of SG in education is also called Game Based Learning (GBL). Following Zyda (2005), GBL activities are designed to help achieve a balance between fun and educational value. GBL could enhance problem-solving competence, decision making, knowledge transfer, and meta-analytical skills (Kirriemuir and McFarlane 2004). Specifically, those games involving collaborative actions can help to put learning into an authentic and realistic context allowing students to practice in a safe environment (Leemkuil et al. 2003). These authors also point to the fact that games can provide realism and motivation to players; they do it through good pedagogical design that brings complexity, risk, role-play, and access to information into the game.

It must be noted that these scenarios may show a lack of effectiveness when no instructional measures or support are added in order to guide this process. In this respect, de Freitas et al. (2010) affirms that negative learning transfer may occur with some game players in SG contexts, where an expectation for high fidelity environments may be related to negative learning processes. Collaborative GBL activities, as a type of Computer Supported Collaborative Learning (CSCL), demand participants to monitor and adapt their cognitive and metacognitive processes, such as temporal competence, to changes in their motivational state (Azevedo 2008). Therefore, we can expect students' TP to play an important role in achievement of optimal learning outcomes in GBL environments. Due to the lack of research in the field of TP in collaborative GBL, we aimed at focusing on analyzing the relation of the students' TP to their game scores (game performance hereinafter).

1.2. Time perspective and learning

This study is based on the definition and operationalization of Time Perspective (TP) by Zimbardo, Keough, and Boyd (1997): "the manner in which individuals, and cultures, partition the flow of human experience into distinct temporal categories of past, present and future". These temporal frames are subdivided into five subscales. Past Negative (PN) individuals are those who present a pessimistic attitude toward the past and possibly the experience of sad events in their past. Past Positive (PP) individuals have a sentimental and positive view of "the old days". Present Hedonistic (PH) will have immediate pleasure, with slight regard to risk without thinking of the consequences, while Present Fatalistic (PF) have no hope for the future and believe that external forces determine their fate. The fifth temporal dimension, the Future (FTP), is characterized by delay of gratification, as a result of the desire of future-oriented individuals to achieve specific long-term goals. An ideal time orientation (high in PP, PH, and FTP) is defined as "balanced" (Zimbardo and Boyd 1999). Individuals with a balanced TP can make plans for the future, consider the past for future successes and possible failures, and enjoy the present.

The importance of TP lies in its relation with different behaviors such as achievement, goal-setting, and risk-taking (Zimbardo and Boyd 1999). TP has been the object of study for educational psychologists because of its relation with learning processes and outcomes. According to Kauffman and Husman (Kauffman and Husman 2004), TP is fundamental in understanding our activities, hopes, goals, and motivations. It was noted that individuals with high Grade Point Average (GPA) are characterized by being future oriented (Mello and Worrell 2006; Ozcetin and Eren 2010). Some authors affirm that college students' thoughts about their future could have an impact on their academic achievement (Shell and Husman 2001). Using a self-report scale instrument, the Temporal Orientation Scale (TOS), Brown and Jones (Brown and Jones 2004) found that past- and present-oriented students were likely to engage in social activities more than academic ones. Future-oriented university students more easily anticipate the implications of their present classroom activities for the distant future (Phalet, Andriessen, and Lens 2004). In a study on TP and academic achievement conducted on African American high school students, Brown and Jones (2004) observed that future-oriented individuals saw education as more useful for future success in life and showed higher GPA.

Education is defined as a future-oriented process because it involves processes oriented toward future goals and delay of gratification (de Bilde, Vansteenkiste, and Lens 2011). Due to this fact, the relation between TP and education has focused on the concept Future Time Perspective (FTP). Nevertheless, GBL as a learning methodology focused on instant rewards, involving competition and social activities (Bateman and Boon 2006), is supposed to help present-oriented individuals to improve their performance and engagement in these activities. Despite a lack of studies in GBL, present-focused individuals perform better in instant feedback situations such as competitions while future-oriented students may engage in seeking academic goals (Kauffman and Husman 2004). There is a need to study how different TP students perform in GBL and explore the possible relation between TP and learning performance.

1.3. Research question and hypotheses
According to the previously conducted experiences in TP and learning, there are empirical and theoretical reasons to affirm that there are no significant differences in a GBL scenario between present-oriented and future-oriented participants. This could be due to two different underlying reasons; based on GBL studies, and as studied by Moreno-Ger et al. (Moreno-Ger et al. 2008), the mix of fun and learning introduced by the GBL methodology could neutralize the heterogeneous learning outcomes expected from the results seen in classic learning activities. Focusing on motivation, present-oriented students prefer instant-reward activities (Wassarman 2002) while future and balanced individuals can foresee investment in learning as a source of future rewards. Therefore, we state two hypotheses: Hypothesis 1 predicts that both individual and collaborative game performance (dependent variable) are not correlated to TP (independent variable); that is, all students can perform equally in a GBL activity. Hypothesis 2 affirms that future-oriented individuals foresee the learning usefulness of the activity in the future, while present-oriented students play for fun, without taking into account the future usefulness of the GBL activity (Hypothesis 2a). On the other hand, as present-oriented students face MetaVals as a game, they may have a similar intention of use in the future as future-oriented. Therefore, all students will have similar intention of use, albeit due to different reasons (Hypothesis 2b).

II. Methods

2.1. Participants
Master students participating in this case study (9 women and 15 men, age M=31.90, SD=4.09, age range: 26–42 years) were engaged in an introductory finance course in ESADE Law and Business school. Names and personal data from participants are treated confidentially and they do not appear in the research. All of the participants in the two expected experiences and the professor were informed of the study and its

purpose. The professional profile of the participants in these programs was composed of marketing and sales, law, and operations experts.

2.2. Research design
To study our hypotheses, the SG MetaVals was developed and implemented in an introductory finance course. The use of a pre-test on finance literacy, together with the GBL activity and a post-test, where students were asked about future usefulness and intentions of use of the game, composed the scenario. All of the activities were set in the Moodle page of the course, and the participants could access the contents one week before the first face-to-face class and one week after. Students played the MetaVals game on their laptops in the context of the first face-to-face class.

2.3. Instruments and operationalization of variables

2.3.1. ZTPI
The analysis of the students' TP was conducted using the Zimbardo Time Perspective Inventory (ZTPI; Zimbardo and Boyd 1999). Fifty-six statements represent the five theoretically independent factors described by Zimbardo and Boyd (Zimbardo and Boyd 1999). Each statement is rated using a 5-point Likert scale (1 = strongly disagree, and 5 = totally agree). After its completion, the ZTPI shows a value of individual's TP. Following these authors, the individuals have a tendency toward one of the five orientations or present a balanced TP. In our research, the participants were found to be present, future, and balanced. The Spanish version of the ZTPI was implemented in Moodle. This instrument had been previously validated through a psychometric study conducted by Díaz-Morales (Díaz-Morales 2006) among a reliable sample of Spanish adults ($N=756$) and was used in the present study to be consistent with the theoretical approach of the chosen TP definition.

2.3.2. MetaVals
MetaVals is a computer-based Serious Game designed by ESADE in the context of the FP7 Network of Excellence Games and Learning Alliance (GaLA). MetaVals was adapted from an existing class activity used to practice basic finance concepts (Massons et al. 2011). Despite the pedagogical interest of the initial activity, only some students actively participated, and it was difficult to incentivize discussion among peers in that context. Therefore, MetaVals was designed through a process that involved a 1) paper-based release, and 2) computer-based versions of the game that were tested in different environments (Padrós, Romero, and Usart 2011). The present MetaVals is a sorting game where students play in dyads with a virtual peer against the rest of the class. A welcome screen asks players to introduce their age and previous knowledge on finance. It leads to a second screen with virtual peers' information (see Figure 1). This key data can help players in the correction and discussion phases (e.g. a virtual peer with a low level on finance may give wrong answers). After general instructions are given by a virtual lecturer, the player starts playing individually by classifying six items as assets or liabilities (e.g. "Computer software", "Bank Loan"); after this first phase, six different items appear, but now the player has access to his virtual peer's answers. After this correction phase, a final discussion phase starts; the player has to decide if the 12 classified items were correctly classified; the dyad with a higher number of correct answers in less time, wins the game.

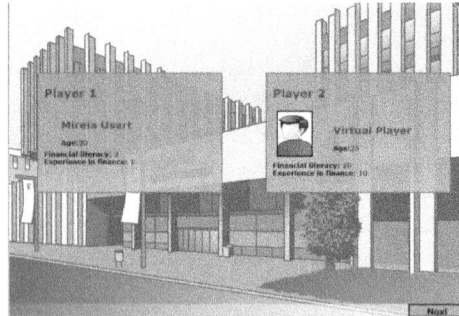

Figure 1. Screen showing the virtual peer's information in the MetaVals Game

The present version of MetaVals implements a countdown in each classifying phase screen and a MySQL database to monitor and record all of the participants' individual and collaborative scores, and time logs. Final scores are an operationalization of the game performance's variable.

2.3.3. Future usefulness and future intentions of use operationalization

After the GBL activity, the students were invited to fill out a questionnaire on future usefulness and future intentions of use for the MetaVals. This instrument was based on the Technology Acceptance Model and had been previously studied in other contexts using MetaVals (Padrós, Romero, and Usart 2011). Four statements on the future uses of the game (3 months and 1 year time) had to be rated by using a 5-point Likert scale (1 = strongly disagree, and 5 = totally agree).

III. Data Analysis and Results

In order to study the two hypotheses, Analysis of Variance or One-Way ANOVA was used. It is important to bear in mind the normality of the sample and equality of variances. Both assumptions were studied. First, the normality test on Origin8Pro (Kolmogorov–Smirnov; K–S) was run for different dependent variables; the use of the K–S test follows the method of different authors on TP that conducted similar experiences (de Bilde, Vansteenkiste, and Lens 2011). It confirmed that the sample followed a normal distribution. The only variables providing an ambiguous result were game performances (both individual and collaborative), but as they were close to the significant level ($p=0.04 <0.05$), we decided to use the parametric test.

Our first hypothesis was aimed at studying whether there was a relation among Time Perspective (TP) and both Individual and Collaborative Game Performance. Participants' scores in the individual phase of the game did not differ significantly across the three TP groups, $F (2, 21) = 0.14$, $p = 0.87$. None of the collaborative scores were significantly different among groups, $F (2, 21) = 2.10$, $p = 0.15$. However, future-oriented showed a higher score for the collaborative phases (M=11.5; SD=0.9) than present (M=10.5; SD=1.51) and future (M=10.5; SD= 1.29) individuals (see Figure 1). Due to the fact that the tendency is not significant, we can confirm the first hypothesis (Figure 2).

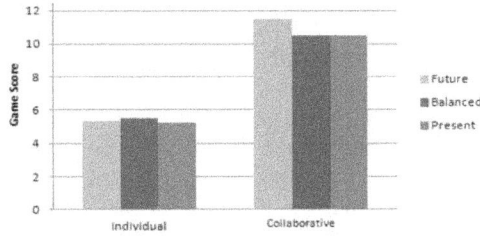

Figure 2. Average scores on the MetaVals Game for the individual and the collaborative phases, comparing the three different TP groups.

For the second hypothesis, students' answers on future Usefulness and Intention of Use in the post-test were analyzed. Results showed that future-oriented individuals believed the game would be useful within one year F (2, 15) = 4.35, p=0.03 (<0.05) when compared to present-oriented participants. This result is significant, and therefore, it confirms Hypothesis 2a. Nevertheless, when asked on future Intention of Use, no significant results showed, although a tendency was clear; future-oriented students made explicit their future intentions of using the MetaVals within less than one year F (2, 15)= 3.21, p=0.07; more studies should be conducted to confirm or reject Hypothesis 2b.

IV. Discussion and Conclusions

The sample was composed of 50% of future-oriented students, 33.33% balanced and 16.7% present-oriented. ANOVA results confirm Hypotheses 1; there is no significant relation between TP and Game Performance, neither for the individual nor for the collaborative phase of the game. Due to the lack of previous studies in the field of Game Based Learning (GBL) and TP, more research should be done to confirm the tendency of future-oriented students to score higher than balanced and present-oriented individuals. This could be faced with a greater sample size and a more difficult game activity that permitted a wider range of scores. Similar results shown among the three TP groups in the game could be confirming the idea that a mix of fun and learning introduced by the GBL methodology (Moreno-Ger et al. 2008) neutralizes different learning performances found in classical learning activities. The underlying reasons for these equal performances could be the fact that present-hedonists tend to engage in instant-reward activities (Wassarman 2002); they face a GBL activity as an amusing, challenging activity. On the contrary, future-oriented students could be engaging in the GBL activity not for fun, but thinking of the learning and future outcomes of playing in an educational context. Finally, balanced individuals adapt their time orientation to the needs of the present moment, having fun and thinking of their future learning gains (Zimbardo and Boyd 1999). The results in the post-test on future usefulness point to this direction; future-oriented participants significantly foresee the usefulness of the game in one-year's time, while present oriented individuals probably play for fun; although they think of playing again, they do not consider the future usefulness of the game (Brown and Jones 2004).

This study could set the groundwork for future research in the field of TP and GBL. Results point to the importance of including GBL activities in management learning courses, which could lead to an equilibrium of performance among students and enhance knowledge acquisition in present-oriented individuals; these goals could be reached by engaging them in activities that give an immediate feedback, such as GBL. These results should also serve as a base for educational psychologists to help individuals in managing their learning processes in terms of performance and usefulness.

4.1. Limitations of the study and future research
One of the limitations of the present study is the size of the sample. The fact that it was the second time that MetaVals was used in a real learning environment could be a handicap. Second, the voluntary filling of the ZTPI questionnaire caused the students to respond in a very irregular number. Researchers cannot generalize the results of the experience; therefore, increasing the size

of the samples, and therefore decreasing the standard deviation is the goal of the researchers for the next month. Concretely, in the context of GaLA, the MetaVals game will be adapted and implemented in Scotland and Romania. The retrieval of data from different samples of adult students may also permit the study of GBL and TP, considering cultural differences between Western and Eastern Europe.

Another limitation of the study is the short time period in which the research was conducted. Following Nurmi (Nurmi 2005), a consequence of the lack of longitudinal data in TP studies is that very little is still known about the antecedents and consequences of TP in learning. In the following months, we will study if GBL performance significantly means an improvement in students' knowledge within a long-term perspective. A longitudinal study of the two masters should help understanding if observed performances are related to deep learning processes and if self-reported future Intention of Use is confirmed in these prospective studies.

About the Authors
Mireia Usart is a junior researcher with the Fellowship at the Direction of Educational Innovation and Academic Quality (DIPQA) at the ESADE School of Law and Business in Spain.
Margarida Romero is the Associate Director of E-Learning at the ESADE School of Law and Business in Spain.

References
Azevedo, R. 2008. "The role of self-regulation in learning about science with hypermedia." In *Recent Innovations in Educational Technology that Facilitate Student Learning*, eds. Robinson, D., Schraw, G., 127-156.

Bateman, C.M. and R. Boon. 2006. *21st Century Game Design*. Hingham, MA: Charles River Media.

Brown, W.T., and J.M. Jones. 2004. "The Substance of Things Hoped for: A Study of the Future Orientation, Minority Status Perceptions, Academic Engagement, and Academic Performance of Black High School Students." *Journal of Black Psychology,* 30 (2): 248-273.

de Bilde, J., M. Vansteenkiste, and W. Lens. 2011. "Understanding the Association Between ftp and Self-regulated Learning Through the Lens of Self-determination Theory." *Learning and Instruction* 21 (3): 332-344.

de Freitas, S., G. Rebolledo-Mendez, F. Liarokapis, G. Magoulas, and A. Poulovassilis. 2010. "Learning as Immersive Experiences: Using the Four Dimensional

Framework for Designing and Evaluating Immersive Learning Experiences in a Virtual World." *British Journal of Educational Technology* 41 (1): 69-85.

Díaz-Morales, J.F. 2006. "Estructura factorial y fiabilidad del Inventario de Perspectiva Temporal de Zimbardo." *Psicothema* 18 (3): 565-571.

Gros, B., E. Barberà, and P. Kirshner. 2010. "Time Factor in e-Learning: Impact Literature Review." *ELearnCenter Research Paper Series*16-31.

Kauffman, D., and J. Husman. 2004. "Effects of Time Perspective on Student Motivation: Introduction to a Special Issue." *Educational Psychology Review* 16 (1): 1-7.

Kirriemuir, J. and A. McFarlane. 2004. "Literature Review in Games and Learning." *Nesta Futurelab series* (8; Bristol).

Leemkuil, H., T. de Jong, R. de Hoog, and N. Christoph. 2003. "KM Quest: A Collaborative Internet-Based Simulation Game." *Simulation & Gaming* 34 (1): 89-111.

Massons, J., M. Romero, M. Usart, S. Mas, A. Padrós, and E. Almirall. 2011. Uso del aprendizaje basado en juegos en la formación de finanzas para no financieros. Actas de las Jornadas Interuniversitarias de Innovación Docente. Universitat Ramon Llull, DEUSTO, ICADE, June 16–17, Barcelona.

Mawdesley, M., G. Long, S. Al-Jibouri, and D. Scott. 2011. "The Enhancement of Simulation Based Learning Exercises through Formalised Reflection, Focus Groups and Group Presentation." *Computers and Education* 56 (1): 44-52.

Mello, Z.R., and F.C. Worrell. 2006. "The Relationship of Time Perspective to Age, Gender, and Academic Achievement among Academically Talented Adolescents." *Journal for the Education of the Gifted* 29 (3): 271-289.

Moreno-Ger, P., D. Burgos, I. Martínez-Ortiz, J. Sierra, and B. Fernández-Manjón. 2008. "Educational Game Design for Online Education." *Computers in Human Behavior* 24: 2530-2540.

Nurmi, J.E. 2005. "Thinking about and Acting upon the Future: Development of Future Orientation Across the Lifespan." In *Understanding Behavior in the Context of Time: Theory, Research, and Application*, eds. A. Strathman and J. Joireman. Mahwah, NJ: Lawrence Erlbaum Associates.

Ozcetin, N. and A. Eren. 2010. "The Effects of Perceived Instrumentality and Future Time Perspective on Students' Graded Performance and Attitudes Regarding English class." *International Journal on New Trends in Education and Their Implications* 1 (4): 5.

Padrós, A., M. Romero, and M Usart. 2011. "Developing Serious Games: Form Face-to-Face to a Computer-based Modality." *E-learning papers* 25: 15-07-2011.

Phalet, K., I. Andriessen, and W. Lens. 2004. "How Future Goals Enhance Motivation and Learning in Multicultural Classrooms." *Educational Psychology Review* 16 (1): 59-89.

Shell, D.F., and J. Husman. 2001. "The Multivariate Dimensionality of Personal Control and Future Time Perspective Beliefs in Achievement and Self-regulation." *Contemporary Educational Psychology* 26: 481-506.

Wall, J., and V. Ahmed. 2008. "Use of a Simulation Game in Delivering Blended Lifelong Learning in the Construction Industry—Opportunities and Challenges." *Computers and Education* 50 (4): 1383-1393.

Wassarman, H.S. 2002. "The Role of Expectancies and Time Perspectives in Gambling Behaviour." *Dissertation Abstracts International: Section B: The Sciences and Engineering* 62 (8B): 3818.

Zimbardo, P.G. and J.N. Boyd. 1999. "Putting Time into Perspective: A Valid, Reliable Individual Differences Metric." *Journal of Personality and Social Psychology* 77: 1271 1288.

Zimbardo, P.G., K.A. Keough, and J.N. Boyd. 1997. "Present Time Perspective as a Predictor of Risky Driving." *Personality and Individual Differences* 23: 1008.

Zyda, M. 2005. "From Visual Simulation to Virtual Reality to Games." *IEEE Computer* 38 (9): 30-34.

www.ingramcontent.com/pod-product-compliance
Lightning Source LLC
LaVergne TN
LVHW081326060426
835511LV00011B/1875